Mechanical Testing for the Biomechanical Engineer

A Practical Guide

Synthesis Lectures on Biomedical Engineering

Editor
John D. Enderle, *University of Connecticut*

Lectures in Biomedical Engineering will be comprised of 75- to 150-page publications on advanced and state-of-the-art topics that span the field of biomedical engineering, from the atom and molecule to large diagnostic equipment. Each lecture covers, for that topic, the fundamental principles in a unified manner, develops underlying concepts needed for sequential material, and progresses to more advanced topics. Computer software and multimedia, when appropriate and available, are included for simulation, computation, visualization and design. The authors selected to write the lectures are leading experts on the subject who have extensive background in theory, application and design.

The series is designed to meet the demands of the 21st century technology and the rapid advancements in the all-encompassing field of biomedical engineering that includes biochemical processes, biomaterials, biomechanics, bioinstrumentation, physiological modeling, biosignal processing, bioinformatics, biocomplexity, medical and molecular imaging, rehabilitation engineering, biomimetic nano-electrokinetics, biosensors, biotechnology, clinical engineering, biomedical devices, drug discovery and delivery systems, tissue engineering, proteomics, functional genomics, and molecular and cellular engineering.

Digital Image Processing for Ophthalmology: Detection of the Optic Nerve Head
Xiaolu Zhu, Rangaraj M. Rangayyan, Anna L. Ells
January 2011

Modeling and Analysis of Shape with Applications in Computer-Aided Diagnosis of Breast Cancer
Denise Guliato, Rangaraj M. Rangayyan
January 2011

Analysis of Oriented Texture with Applications to the Detection of Architectural Distortion in Mammograms
Fábio J. Ayres, Rangaraj M. Rangayyan, J. E. Leo Desautels
2010

Fundamentals of Biomedical Transport Processes
Gerald E. Miller
2010

Models of Horizontal Eye Movements, Part II: A 3rd Order Linear Saccade Model
John D. Enderle , Wei Zhou
2010

Models of Horizontal Eye Movements, Part I: Early Models of Saccades and Smooth Pursuit
John D. Enderle
2010

The Graph Theoretical Approach in Brain Functional Networks: Theory and Applications
Fabrizio De Vico Fallani, Fabio Babiloni
2010

Biomedical Technology Assessment: The 3Q Method
Phillip Weinfurt
2010

Strategic Health Technology Incorporation
Binseng Wang
2009

Phonocardiography Signal Processing
Abbas K. Abbas , Rasha Bassam
2009

Mechanical Testing for the Biomechanical Engineer: A Practical Guide
Marnie M. Saunders

ISBN: 978-3-031-00534-3 print
ISBN: 978-3-031-01662-2 ebook

DOI: 10.1007/978-3-031-01662-2

A Publication in the Springer series
SYNTHESIS LECTURES ON ADVANCES IN AUTOMOTIVE TECHNOLOGY
Series Editor: John D. Enderle, University of Connecticut

Series ISSN 1930-0328 Print 1930-0336 Electronic

Mechanical Testing for the Biomechanical Engineer

A Practical Guide

Marnie M. Saunders

The University of Akron, Akron, Ohio

SYNTHESIS LECTURES ON BIOMEDICAL ENGINEERING #54

ABSTRACT

Mechanical testing is a useful tool in the field of biomechanics. Classic biomechanics employs mechanical testing for a variety of purposes. For instance, testing may be used to determine the mechanical properties of bone under a variety of loading modes and various conditions including age and disease state. In addition, testing may be used to assess fracture fixation procedures to justify clinical approaches. Mechanical testing may also be used to test implants and biomaterials to determine mechanical strength and appropriateness for clinical purposes. While the information from a mechanical test will vary, there are basics that need to be understood to properly conduct mechanical testing. This book will attempt to provide the reader not only with the basic theory of conducting mechanical testing, but will also focus on providing practical insights and examples.

KEYWORDS

biomechanics, orthopaedics, mechanical testing

Contents

Preface

This book is intended to provide an introduction to the practical aspects of conducting biomechanical testing. As such it targets the biomechanical engineering student with a basic understanding of solid mechanics but with little practical experience.

The take-home message of this book is provided on the first page. In order to obtain reliable data from mechanical testing it is critical to take the time to conduct an accurate and reliable test. Ensuring that the test is conducted properly entails a conscious effort such that one is: understanding machine and testing basics; properly setting up tests by correctly using measurement tools; understanding how to load a specimen to get physiologically relevant data; designing/selecting appropriate testing fixtures; properly running the testing machine; and properly analyzing the data. The goal of any test is to obtain reliable data that can be confidently used to determine the study conclusions. The attention to detail in testing design, setup, running and analysis is key to obtaining reliable data. Here we discuss these details in the context of practical applications and examples.

Acknowledgments

I have been fortunate to work with wonderful engineers, surgeons, and machinists over the last 20 years. There are three that have been true heroes to me in every sense of the word and I would like to dedicate this work to them.

Dr. Daniel Sheffer, former Chair of the Department of Biomedical Engineering, The University of Akron: His tireless dedication to the department and a true selflessness in building the undergraduate and graduate programs was inspiring. His love of teaching was infectious and his kindness and support have resonated throughout my life.

The late Dr. Glen Njus, former Associate Professor in the Department of Biomedical Engineering, The University of Akron: I began working for Dr. Njus as an undergraduate in Mechanical Engineering. He gave me the space to make mistakes and the belief in myself that I could be a productive engineer. As my graduate advisor, he gave me the freedom to pursue my own research interests and shared with me a fascination and love of orthopaedic biomechanics. With his untimely death, I had the opportunity to return to my alma mater. While his shoes are too big for me to fill, I will strive to do my best to continue the orthopaedic biomechanics program he built.

Larry Saunders, a machinist for over 50 years: My father taught me to value the practical aspects of machines and engineering; the ability to fabricate rather than merely design. My love of engineering started early when, as the youngest of 2 daughters, I found myself as the mechanic's helper on countless weekends. I carried the toolbox, handed the tools, ran any errands, and held the light. Although more often than not in the way, I eagerly anticipated every weekend. He could smell a screwdriver sale just by walking into Sears and I have him to thank for my colorful vocabulary. Above all, I thank him for his love and support.

CHAPTER 1

Fundamentals

Mechanical testing is part science and part art. Mechanical testing requires a rigorous understanding of basic mechanics, mathematics and physics, in addition to the ability to design fixtures and protocols that will result in accurate, reproducible and reliable data. For example, Figure 1.1 depicts a mechanical loading platform outfitted with a pair of friction grips being used to determine the mechanical strength of a rodent skin section, post-wound healing. The science of the testing combined with the art of the design and fabrication of the fixtures results in accurate and reliable data by which to assess healing strength. Here we will attempt to introduce the basics of both aspects of mechanical testing. We will begin by reviewing basic mechanics relevant to mechanical testing. We will then discuss fundamental aspects of mechanical testing including measurement and measurement tools, design and machine development. We will then provide examples and practical insights into accomplishing mechanical testing related to fixture design and testing application. Finally we will provide detailed descriptions of a few projects representing different aspects of orthopaedic biomechanics and address study details not generally discussed in published articles.

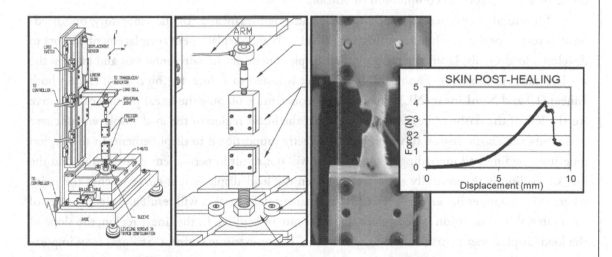

Figure 1.1: Mechanical testing is part science and part art. Here friction grips have been designed and machined to hold a soft tissue strip in place during a tension test. The data collected during the test is used to quantifiably assess the mechanical performance of the "healed" skin segment.

1.1   BASIC MECHANICS

1.1.1   MECHANICAL PROPERTIES

When an object, such as a bone or implant, is under mechanical load its performance is governed by its size, shape and the material that comprises it. Collectively, these features determine the mechanical properties owing to the fact that they govern the mechanical performance of the object. Mechanical properties are of two types: material properties and structural properties. Material properties are independent of geometry and represent the substance which makes up the specimen. Structural properties are dependent upon geometry and represent the behavior of the structure as a function of the size and shape, or form, it takes. The difference between these concepts can be easily understood if thought of in terms of extremes. For example, in Figure 1.2, the steel I-beam (structural building support) and the steel bolt have identical material properties because they are both made from steel. However because the I-beam is several times larger than the bolt, the structural properties will be quite different. Similarly, the stainless steel dental screw and the stainless steel long bone screw have identical material properties but different structural properties, while the ceramic femoral head and the stainless steel femoral head have different material properties. Mechanical properties are determined by mechanical testing and are associated with a particular mode of loading, such as compression or tension.

Structural properties are experimentally measured quantities. In the laboratory, one needs only to test a specimen while recording load (load cell) and displacement (displacement sensor) to develop a load vs. displacement curve plotting displacement on the horizontal axis and load on the vertical axis. As shown in Figure 1.3, if an object is tested to failure, the object will exhibit both Linear (L) and NonLinear (NL) behavior. If we concentrate on only the linear portion of the curve (to the left of the dashed line), it can be seen that the linear region of the load-displacement curve represents the elastic region in which load is directly proportional to displacement. In this region repetitive loading and unloading of the material will not result in permanent deformation, and the rate of loading is theoretically not a factor. When the load-displacement curve reaches the point where it is no longer linear, it has "yielded," permanent deformation will result and the behavior of the material in this region is dependent upon the rate of loading. In the linear region, the slope of the load-displacement curve is called the "stiffness" (k), or extrinsic stiffness. Stiffness is an important structural property and is a measure of the compliance of the specimen. That is, the greater the stiffness of an object, the less likely it is to deflect or deform under a given load.

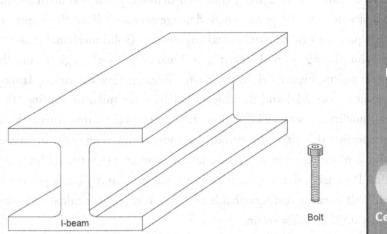
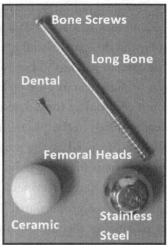

Figure 1.2: (left) The steel I-beam and bolt illustrate the concept of material and structural properties. If both objects are made from steel, they will have the same material properties. However, given the difference in size and geometry, they will not have the same structural properties. (right) The dental and orthopaedic stainless steel screws would have the same material properties, while the femoral heads (ceramic and stainless steel) with similar geometries would not have the same material properties.

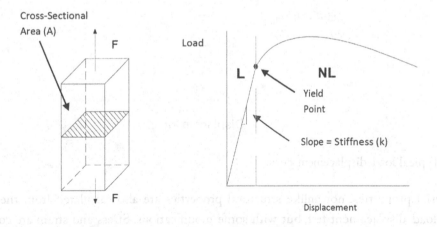

Figure 1.3: Structural properties are determined from mechanical testing. If a tensile load is applied to the block, the block will undergo an elongation, or a change in length with the applied load. If the load and displacement are recorded, the load-displacement curve will represent the mechanical behavior of the object. If tested to failure, the specimen will experience both elastic and plastic deformation. The linear (L) portion of the curve represents the area in which loading and subsequent unloading will not cause permanent deformation (damage) to the object. The nonlinear (NL) portion of the curve represents the area in which loading and subsequent unloading will cause permanent deformation to and eventual failure of the object.

The yield point represents a somewhat imaginary (not well-defined) point that distinguishes the change in the loading of a structure to the point where damage accrues. Given the import of this point in design, depending upon the field of mechanical engineering (solid mechanics, fracture and fatigue, metallurgy), this point distinguishes the elastic, or linear or pre-yield region from the plastic, or nonlinear or post-yield region, Figure 1.4. In the elastic (linear, pre-yield) region, damage does not accumulate (or is assumed reversible) and the object's behavior is similar to a spring (theoretically, F=kx). In the plastic (nonlinear, post-yield) region, damage is accumulated. Theoretically the yield is preceded by the proportional limit at the end of the linear region. Generally in biomechanics research the yield point is more relevant and a distinction between proportional limit and yield point is often not evident. If a distinction between proportional limit and yield is present in the specimen the proportional limit is easily distinguishable and the yield point is calculated using offset techniques that are usually 0.1% to 0.2% strain.

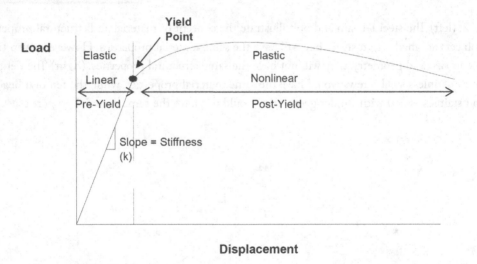

Figure 1.4: Typical load-displacement curve.

Material properties, not unlike structural properties are also calculated from the data obtained in a load-displacement test, but with some modifications. Stress and strain are constitutive relations and neither can exist independent of the other. For a tension test, "stress" is simply a measure for a force (load) over a given area (F/A) and "strain" is a measure of the extent to which an object changes in length with an applied load and measured with respect to its initial length ($\Delta L/L_o$). That is, the loads in the load-displacement data divided by the cross sectional area of the object will generate the stresses. In addition, the incremental displacement data taken with respect to the initial length of the specimen will generate the strains (Figure 1.5). A consequence of this simple conversion is that the load-displacement and stress-strain curves will have identical shapes, but different numerical scales. This is illustrated qualitatively in Figure 1.5.

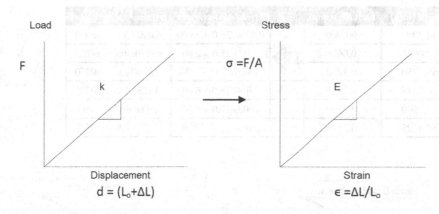

Figure 1.5: Mechanically loading an object in the elastic region yields a linear load-displacement curve. The slope of this curve is the structural property known as stiffness (k). While the structural properties can be determined from this curve, material properties may be determined by converting the load-displacement curve into a stress-strain curve by taking into account the object dimensions. The slope of this curve is the material property known as Young's (Elastic) Modulus (E). Since A and L_o are constant, the curves are identical in shape but vary in magnitude.

An important concept to grasp in mechanical testing is that mechanical properties are determined from a single test. For example, if a tension test is run, the structural properties are determined from the load-displacement curve and values such as the extrinsic stiffness (k) may be calculated. To obtain the material properties the same data is manipulated such that the geometric dependence is removed and the resulting plot is the stress-strain curve which reflects material behavior. Values such as the intrinsic stiffness (Young's Modulus (E)) may be calculated from the stress-strain curve. This is illustrated in the example below.

Example (Calculating Young's Modulus)

The specimen shown in Figure 1.6a has an initial length of 50.0000 mm with a cross section of 5.00 mm by 5.00 mm (A = 25.0 mm^2). The object is placed in a testing machine where it is subjected to a tensile load causing it to elongate. As shown in the table, the load increases from 0.0000 to 250.0 N and displacement is recorded. Plotting the load-displacement curve we obtain Figure 1.6b. Converting to stress-strain and plotting we get Figure 1.6c.

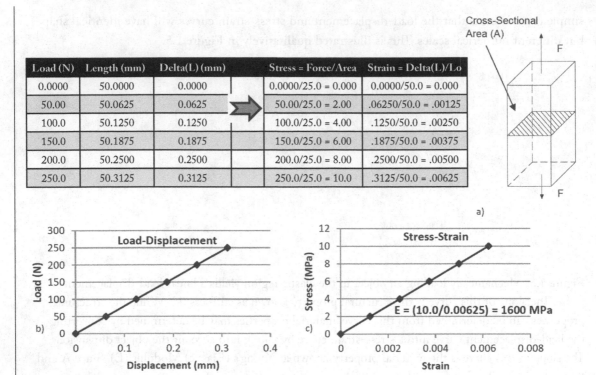

Load (N)	Length (mm)	Delta(L) (mm)		Stress = Force/Area	Strain = Delta(L)/Lo
0.0000	50.0000	0.0000		0.0000/25.0 = 0.000	0.0000/50.0 = 0.000
50.00	50.0625	0.0625		50.00/25.0 = 2.00	.06250/50.0 = .00125
100.0	50.1250	0.1250		100.0/25.0 = 4.00	.1250/50.0 = .00250
150.0	50.1875	0.1875		150.0/25.0 = 6.00	.1875/50.0 = .00375
200.0	50.2500	0.2500		200.0/25.0 = 8.00	.2500/50.0 = .00500
250.0	50.3125	0.3125		250.0/25.0 = 10.0	.3125/50.0 = .00625

Figure 1.6: Calculating Young's Modulus from a tension test.

Given the linearity of the curves, Young's Modulus may be calculated from any two points on the curve as the slope of the linear region or the change in the rise over the change in the run (Δrise/Δrun) of the stress-strain curve. For example, between the extremes (0, 0.000) and (10.0, .00625) a Young's Modulus of 1600 MPa is calculated; similarly, between (4.00, .00250) and (8.00, .00500) a Young's Modulus of 1600 MPa is calculated.

Instead of plotting the load-displacement curve, converting to stress-strain and calculating the slope of the stress-strain curve in the linear region to determine Young's Modulus, one could also work from the fundamental relationships represented by the curves to arrive at the formula for Young's Modulus for a tensile test:

$$\text{Stress } (\sigma) = \text{Force/Area} = F/A \qquad (i)$$
$$\text{Strain } (\varepsilon) = \text{Delta } (L)/L_o = \Delta L/L_o \qquad (ii)$$
$$\text{Young's or Elastic Modulus } (E) = \sigma/\varepsilon \qquad (iii)$$
$$\text{and, therefore: } \sigma = E\varepsilon \qquad (iv)$$

Setting (i) and (ii) into (iv) yields:

$$F/A = E\,(\Delta L/L_o) \qquad (v)$$

And on rearranging:

$$E = F/A * (L_o /\Delta L) \qquad\qquad (vi)$$

Substituting the problem parameters into equation (vi) yields:

$$250.0/25.0 * (50.00/.3125) = 1600 \text{ MPa}$$

The above example illustrates how Young's Modulus may be determined in two ways. First, it can be determined from the conversion of the load-displacement data to stress-strain data and then calculated as the slope of the stress-strain curve in the linear region. Second, it can be obtained from manipulation of the governing equations relating stress and strain to load and displacement, respectively. Later in the book we will provide the fundamental relations for more complex testing scenarios, specifically bending and torsion.

Additional material properties that are obtained from the stress-strain curve are shown plotted in Figure 1.7. Stresses are associated with the strength of the material. As such, the stress at yield is also referred to as the yield stress or "yield strength." The largest stress an object can withstand is the ultimate stress or "ultimate strength" and the stress at failure is also known as the "failure strength" or "breaking strength." As shown in Figure 1.7, for the stress-strain behavior of this object, the ultimate and breaking strength are not equivalent. This is not always the case.

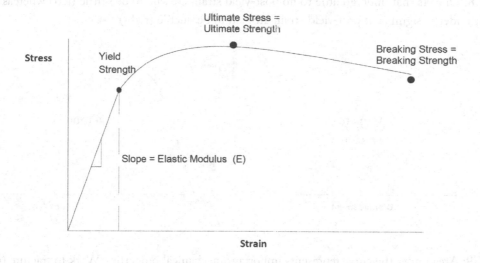

Figure 1.7: Typical stress-strain curve.

Post-yield behavior represents an important part of material behavior. That is, the amount of post-yield strain that a material can endure prior to failure is a measure of the ductility of the material. Materials that are "ductile" undergo a large amount of post-yield deformation and therefore do not fail catastrophically. Materials that do not undergo large post-yield strain are said to be "brittle" and fail catastrophically, or without warning, Figure 1.8.

While the ductility or brittleness of a material deals only with the post-yield portion of the stress-strain curve, the total area under the stress-strain curve (pre-yield + post-yield) is a measure of the energy absorbed to failure, or the "toughness." The equivalent region or the total area under the load-displacement curve (pre-yield + post-yield) represents "work to fracture," Figure 1.9.

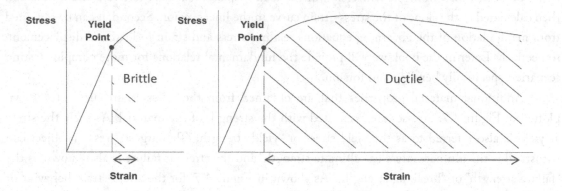

Figure 1.8: Objects that undergo little to no post-yield strain are said to be brittle (left) whereas objects that undergo significant post-yield strain are said to be ductile (right).

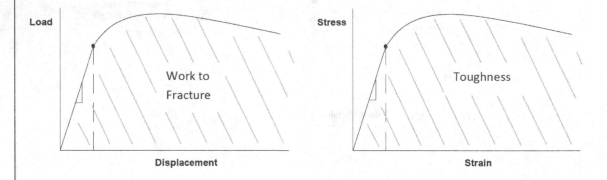

Figure 1.9: Area under the curve represents important mechanical properties. Work to fracture (energy absorbed to failure) is represented by the area under the load-displacement curve (left). Toughness is represented by the area under the stress-strain curve (right).

1.1.2 LOADING MODES

Objects can be subjected to a variety of different loading modes, as demonstrated in Figure 1.10. Here a specimen is shown before and after loading in compression, tension and shear, respectively. While tension and compression occur along the same axis perpendicular to the loading plane, they have opposite effects on the shape of the material; both represent normal stresses. The shear occurs parallel to the plane and has the effect of moving part of the object relative to another part. Stresses can be succinctly defined by the affect they have upon an object. For example, dilatational (e.g., hydrostatic) stresses (Figure 1.10) result in volume changes whereas deviatoric (e.g., shear) stresses result in volume distortion.

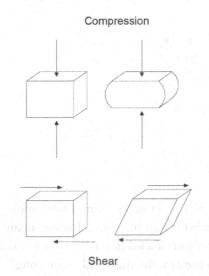
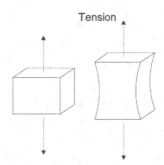

Figure 1.10: Objects may be loaded in different loading modes. Three of those modes, compression, tension, and shear, are illustrated here. As shown, compression has the tendency to shorten and bulge an object; tension has the tendency to lengthen and thin an object; and shear has the tendency to distort an object.

The mechanical loading mode (e.g., compression) is not uniform throughout an object. This implies that if an object is subjected to a compressive load, every point within that specimen is not in compression. This may at first seem confusing but it is an important concept to grasp when studying mechanical loading effects on objects. Even the simplest of loading modes results in complex stress and strain states throughout the object. As such, compression on a uniform, cubed specimen, for example a bone coupon or scaffold material can result in regions of the specimen experiencing shear. Given that bone, like many objects is weaker in shear than in compression a uniform compression load causes the bone to fail in its weakest loading mode, shear. This is illustrated in Figure 1.11. The shear experienced is highest along a plane 45° to the normal, and fracture occurs along this plane. This is also the reason that a long bone subjected to pure torsion will experience a spiral fracture. This behavior is not unique to bone and this effect can be easily demonstrated by applying pure torsion to a piece of blackboard chalk or a carrot. Furthermore, this demonstrates that by understanding how the various mechanical loading modes affect bone, one can look at a bone fracture to infer the type of loading that caused the failure. This is not limited to the loading mode, but is also reflected in the speed with which the fracture occurred. For example, when bone is fractured slowly, such as occurs during low energy trauma (falling from a tree), the bone has time to absorb energy, and a stable, noncomminuted fracture results. When bone is fractured quickly, such as occurs during high energy trauma (a motor vehicle accident or gunshot wound), the bone does not have time to absorb the energy, and the fracture is comminuted, or consists of several fragments.

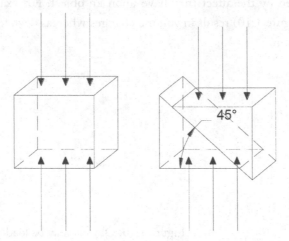

Figure 1.11: Objects under simple loading can experience complex stress states. Even if this simple object is loaded in pure compression, shear stress will be generated within the object. Because many objects, including bone, are weaker in shear than compression, objects loaded in this manner will fail by shear. Given that shear is maximal on a plane 45 degrees to the axis, the shear will occur along this plane, as demonstrated.

1.1.3   MATERIAL PROPERTIES AND DEGREE OF ANISOTROPY

The slope of the stress-strain curve in the linear region is known as the modulus of elasticity or Young's Modulus (E) (Figure 1.7) and is one of three material properties that are needed to uniquely define a simple, homogenous material. The material property known as Poisson's ratio (v) is a measure of how elongating or compressing an object will cause it to neck or bulge, respectively. The Poisson's ratio is a measure of how loading in the longitudinal direction (axially) affects the structure transversely (laterally). With few exception (auxetics) axial tension results in transverse necking (contraction) while axial compression results in transverse bulging (Figure 1.10). For practical purposes the formulation for Poisson's ratio (- lateral strain/axial strain) is preceded by a negative to account for this inverse relationship and yield a positive numerical value. For many biomedically relevant materials (bones, biocompatible metals), a reasonable value of Poisson's ratio is between 0.3 and 0.5, with 0.5 representing incompressible materials. The third material property, the shear modulus (G) is the slope of the shear stress-shear strain curve in the linear region. The data obtained from the torsion test is plotted in a torque-twist curve. Following conversion to the shear stress-shear strain curve, the slope in the linear region represents G, Figure 1.12.

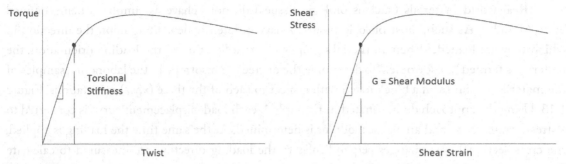

Figure 1.12: Data collected in a torsion test is represented on a torque-twist curve. Material property performance under torsion is represented by the shear stress-shear strain curve.

While the three material properties define a material, the three properties are not independent. That is, determination of any two can enable the calculation of the third, manipulating the relation,

$$G=E/[2(1+v)].$$

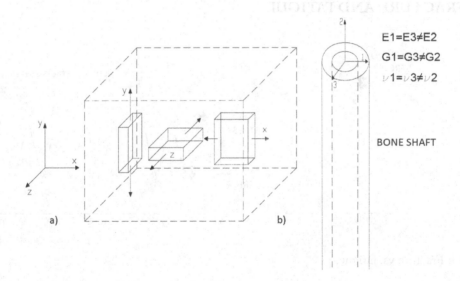

Figure 1.13: Material properties often have directional dependence. When material properties are a function of the loading orientation, this degree of anisotropy can be determined by cutting samples from a section of the object and loading these identical samples in each of the three orientations. Bone is often classified as transversely isotropic implying that the bone behaves the same in the transverse plane, but not in the longitudinal plane, as shown for a cylindrically shaped bone shaft, where E1=E3, G1=G3 and v1=v3.

Real-world materials (such as biologic tissues) do not behave as simply as being defined by E, G and ν. As such, most biologic tissues behave differently depending upon the direction in which they are loaded. When a material's properties are a function of the loading orientation, the material is termed "anisotropic." To determine the degree of anisotropy in the laboratory, samples of the material are cut from a block tested with respect to each of the three (x,y,z) orientations, Figure 1.13. Using the approach demonstrated in Example 1, each load-displacement curve is converted to a stress-strain curve, and an elastic modulus is determined. At the same time the loading is applied, the cross sectional area changes perpendicular to the loading direction are measured to calculate Poisson's ratio. If the moduli and Poisson's ratios for each of the three orientations are the same, the material is said to be "isotropic" and the material is defined by one E, one G and one ν (3 terms). If the values for each of the three orientations are not the same, the material is said to be "ortho-tropic" and defined by three E's, three G's and three ν's (9 terms). Some materials, such as bone, are considered transversely isotropic (6 terms). That is they are assumed to behave roughly the same in a plane transverse to the long axis of the bone but behave differently along the long axis, Figure 1.13.

1.1.4   FRACTURE AND FATIGUE

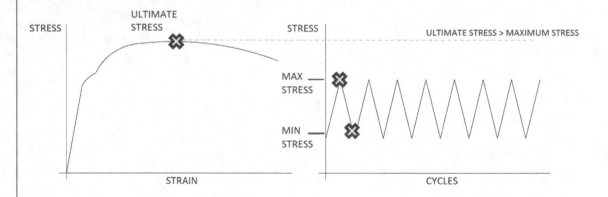

Figure 1.14: Fracture vs. fatigue.

Mechanical tests in which a specimen is loaded to failure can be classified as fracture or fatigue. Although fracture is the term given to failure or rupture of the object, a fracture test is a test in which an object is loaded to failure in a single loading application. In contrast, a fatigue test is conducted when an object is loaded to failure in a multiple loading application (2 cycles +). As such, fatigue tests take place at loads below the fracture load, Figure 1.14. Fatigue is most often represented by an S-N curve where stress (S) is plotted against the number of cycles (N). Cycles are commonly plotted logarithmically. As the stress is reduced, the number of cycles that the object can undergo prior to failure increases. The endurance limit is an ideal limit on the S-N curve, Figure 1.15. Theoretically, objects fatigued at stresses below the endurance limit can undergo an infinite number of cycles without failure. Given that everything eventually fails, the endurance limit is not real, but an idealized limit. However, it is an important design parameter to consider when designing plates/constructs/biomaterials that are exposed to significant cyclic in-service loading.

Selecting the amplitude waveform for a fatigue test is an important consideration. The R-ratio is a ratio of the minimum stress to the maximum stress that a specimen is cycled between to failure. For biomechanics tests, it is advantageous to quantify the fatigue behavior in a physiologically relevant simulation. For example, in a fully reversible fatigue test, (R=-1), the object is subjected to equal magnitudes of tensile and compressive stresses, Figure 1.16. While this may be very appropriate to test the performance of a novel biomaterial, if a synthetic tendon/ligament that experiences tensile in-service loads is fatigued this waveform may not be appropriate. In the case of the tendon/ligament an R-ratio between 0 and 1 would be more appropriate. For an R-ratio of 0, the load is completely removed with each cycle, whereas with anything greater than 0 the implant will experience some level of tensile stress at all times during the test.

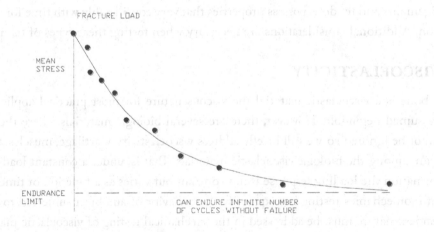

Figure 1.15: Typical S-N curve from fatigue testing. Each point on the curve represents an object (material, implant, construct) tested to failure in fatigue at a given stress with cycles recorded. Here, 11 objects were fatigued with the first point representing the fracture load obtained in a single loading application.

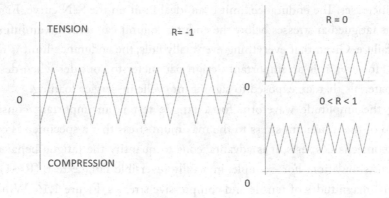

Figure 1.16: Fatigue loads are represented by R-ratios defined as the minimum stress/maximum stress. The plots represent stress vs time.

Much of the previous discussion has centered around relatively rigid objects such as bone or implant metals. These materials are relatively easy to work with because their behavior does not vary considerably as a function of time for a given load or displacement. In contrast, structures such as muscles, ligaments and tendons possess properties that vary considerably with time for a given load/deformation. Additional considerations are necessary when testing these types of tissues.

1.1.5   VISCOELASTICITY

Although bone is a viscoelastic material the viscous nature for most practical applications is ignored or assumed negligible. However, there are several biologic materials where the viscoelastic nature cannot be ignored so we will briefly address viscoelasticity. Cartilage, muscles, tendons and ligaments are among the biologic viscoelastic materials. That is, under a constant load or displacement/deformation the loading response is not constant but varies as a function of time. Given that the goal of biomechanics testing is to quantify the behavior of an object subjected to loading, the time-dependent nature must be addressed in the mechanical testing of viscoelastic materials.

During mechanical testing loading machines may be operated under load or displacement control. For viscoelastic materials a constant load control test is known as a "creep" test, while a constant displacement control test is known as a "stress relaxation" test. The input is shown for a creep test with the corresponding displacement (strain) response in Figure 1.17. The input is shown for a stress relaxation test with the corresponding load (stress) response in Figure 1.18.

INPUT CURVE/ CONSTANT STRESS   OUTPUT CURVE/STRAIN RESPONSE

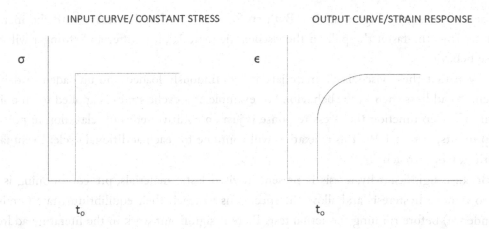

Figure 1.17: In conducting a creep test (in load control), a load is applied to the specimen and held constant with time. The strain response is recorded as a function of time. Initially there is a quick, linear elongation, followed by a gradual, non-linear elongation that reaches a constant (equilibrium).

INPUT CURVE/ CONSTANT STRAIN   OUTPUT CURVE/STRESS RESPONSE

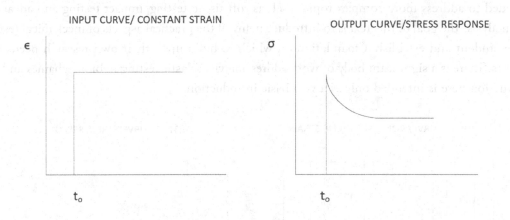

Figure 1.18: In conducting a stress relaxation test (in displacement control), a displacement is applied to the specimen and held constant with time. The stress response (relaxation) is recorded as a function of time. Initially there is a quick, linear stress that decays nonlinearly until it reaches a constant load (equilibrium).

As was previously noted, elastic materials (in the linear loading region) behave as a spring. When loaded, they exhibit a linear response with the stiffness representing the spring constant. Theoretically, the response is not dependent upon the loading rate and the object will achieve the same load for a given displacement. Furthermore, when the load is removed the material returns to its initial, undeformed state. For viscoelastic materials this is not the case. The load is dependent upon the rate at which the strain is applied. There will be an initial peak load that is dependent upon the rate of strain (higher strain). Both stiffness and load will increase with the increasing strain rate. Thus the faster the pull on the viscoelastic material, the stiffer and stronger will be the response behavior.

Given that these viscoelastic materials are continuously loaded and unloaded in daily use, it is useful to address their cyclic behavior. For example, if a cyclic strain is applied to an object in the form of a step function the stress response is just an additive series of relaxation responses for the step inputs, Figure 1.19. This relaxation will continue for each additional cycle. Eventually an equilibrium will be reached.

Because significant hysteresis is present in viscoelastic materials, pre-conditioning is completed to remove hysteresis and allow the specimens to reach their equilibrium state (strain rate independence) before running the actual test. There is significant work in the literature addressing the loading magnitude and cycle number necessary for pre-conditioning various tissues. It is important to note that the purpose of this book is to serve as an introductory practical guide for biomechanical testing. As such, much of what is presented here are the basics and additional reading is needed to address more complex topics such as soft tissue testing, impact testing and advanced data analysis. The goal of this text is to introduce many of the practical aspects of mechanical testing to the student and establish a foundation on which to build upon their own research needs and interests. There is a significant body of work addressing viscoelastic testing in biomechanics and the information here is intended only as a very basic introduction.

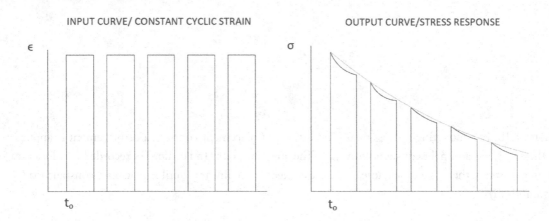

Figure 1.19: In conducting a stress relaxation test (in displacement control), a displacement is applied to the specimen and held constant with time. The stress response (relaxation) is recorded as a function of time. Initially there is a quick, linear stress that decays nonlinearly until it reaches a constant load (equilibrium).

1.1.6 COMPLEX STRESS STATES

As noted earlier in this chapter, even in the simplest loading scenarios, complex stress states can arise internally within the structure. For the simple object shown in Figure 1.20, the axial load applied (tension) to the surface results in shear stresses arising internally on planes at an angle to the original loading direction. The take-home message of this example is that shear stresses arise in all objects loaded; even when subjected to pure axial loading such as the case of simple tension.

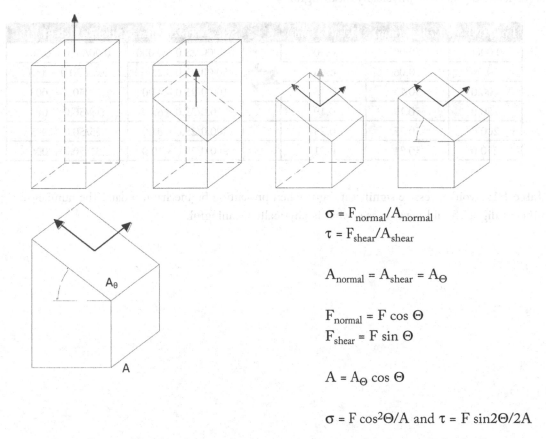

$$\sigma = F_{normal}/A_{normal}$$
$$\tau = F_{shear}/A_{shear}$$

$$A_{normal} = A_{shear} = A_{\Theta}$$

$$F_{normal} = F \cos \Theta$$
$$F_{shear} = F \sin \Theta$$

$$A = A_{\Theta} \cos \Theta$$

$$\sigma = F \cos^2\Theta/A \text{ and } \tau = F \sin2\Theta/2A$$

Figure 1.20: Even the simplest of loading modes (tension/compression) induce complex loading states including shear stresses in simple, homogenous structures of uniform geometry.

1.1.7 SIGNIFICANT DIGITS

Finally, in Example 1, the numbers were chosen for illustrative purposes and mathematical convenience. It is important to remember that even though the equipment may read out to several decimal places, it is rarely appropriate to assume this level of accuracy in testing. Furthermore, the equations used to obtain the values of extrinsic and intrinsic stiffness are not "laws" (for example Hooke's Law) in the formal sense, but merely experimental observations that hold up relatively well to a first-order approximation. In this case it is usually more than sufficient to take biomechanics data out to three to four significant digits when reporting final results. That is, the example for the calculation of Young's Modulus in this chapter was for illustrative purposes and not correct with respect to the data being expressed in significant figures, Table 1.1. Never take data out to more significant digits than physically meaningful.

Load (N)	Length (mm)	Delta(L) (mm)		Stress = Force/Area	Strain = Delta(L)/Lo
0.0000	50.00	0.00		0.0000/25.0 = 0.000	0.00/50.0 = 0.000
50.00	50.06	0.06	→	50.00/25.0 = 2.00	.063/50.0 = .001
100.0	50.13	0.13		100.0/25.0 = 4.00	.13/50.0 = .003
150.0	50.19	0.19		150.0/25.0 = 6.00	.19/50.0 = .004
200.0	50.25	0.25		200.0/25.0 = 8.00	.25/50.0 = .005
250.0	50.31	0.31		250.0/25.0 = 10.0	.31/50.0 = .006

Table 1.1: Avoid excessive significant digits when presenting biomechanical data. The number of significant digits should never exceed what is physically meaningful.

CHAPTER 2

Accuracy and Measurement Tools

2.1 ACCURACY AND PRECISION

Before discussing setting up and conducting biomechanics studies, it is important to understand how to use measurement tools in the laboratory to increase the accuracy and precision of the work. Accuracy is a term used to describe how close to the true value a measurement is; the more accurate the measurement, the closer it is to the true value (which may or may not be known). Precision is a term used to describe how close repeated measurements are to each other, independent of accuracy. The more precise the measurements, the closer they are to each other but not necessarily the true value. As such, measurements (and the equipment used to obtain them) will fit into one of four categories: a) neither accurate nor precise; b) precise but not accurate; c) accurate but not precise; or, d) accurate and precise. A common way to visualize these combinations is demonstrated in Figure 2.1 with the use of a dartboard. The goal is to throw as many darts at the target (center). In (a), the darts (indicated by yellow dots) are neither accurate nor precise. That is, they are not near the center target (accurate) nor clustered together (precise). In (b), the darts are precise with repeated throws lying very close to each other, but not accurate and far from the center of the dartboard. In (c), the darts are accurate in that they are near the center target, but not precise with repeated throws not clustered together. In (d), the darts are both accurate and precise. The accuracy and precision of measurements are dependent upon two things, the quality of the measurement tools (and proper care and maintenance) and the ability to correctly use them.

Figure 2.1: A dartboard is used to represent the concepts of accuracy and precision. The darts (represented by the yellow dots) are: (a) neither accurate nor precise; (b) precise but not accurate; (c) accurate but not precise; and, (d) both accurate and precise.

2.2   MEASUREMENT TOOLS

One of the most important skills that an engineer working in a testing laboratory can gain is the ability to accurately measure. In previous decades the engineering and architectural scales were explained to students in manual drafting classes. With the increasing prevalence of computer programs and solid modeling classes replacing manual drafting classes, the ability to read an architectural or an engineering scale is not nearly as necessary or as commonplace as it once was. However, it is an absolute necessity that the biomechanical engineer learn to use the measuring tools found in a biomechanics laboratory/machine shop. These include the steel rule, dial caliper, micrometer, and Vernier scales.

2.2.1   STEEL RULE

First, it is necessary to distinguish between tools using the International System of Units (SI) (metric) and English units. Before using any measuring tool always check to make sure that you know the unit system you are using. While most countries have adopted the metric system, and biomechanics work is published using the metric system, here in the U.S. it is still quite commonplace to find machinists working in the English system and to purchase equipment using English units. In the machine shop and biomechanics laboratory, measurement devices range from the simple steel rule (generally in 1/32 or 1/64 inch graduations) to the precision-measuring instruments such as calipers, micrometers and Vernier tools (generally in 0.001 (1/1000 inch) to 0.0001 (1/10000 inch) graduations). Shown (Figure 2.2) is a steel rule with combination measurements in metric and English (left) and a steel rule in both 1/8 inch and 1/16 inch graduations (right).

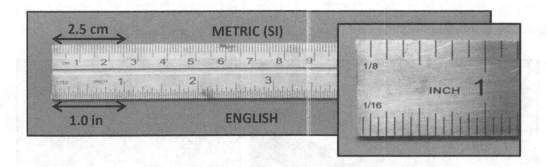

Figure 2.2: The steel rule may be used when only minimal accuracy is necessary. The precision of the tool will depend upon the graduation and at best can only be approximated to halfway between two graduations.

In general it is good practice to avoid taking measurements from the end of the rule as with use the ends have a tendency to become rounded. It is preferable to make measurements from a major graduation, such as from the 1 inch mark. As shown in Figure 2.3, a measurement of 3 7/8 inches is marked on the rule, which has 1/8 inch graduations. The measurement of 2 7/16 inches on the second rule illustrates that the measurement cannot be read to any higher level of precision (resolution) than to one-half the distance represented by two divisions. In this case, the distance between divisions is 1/8 inch, and the measurement can be approximated to one-half of that value or 1/16 inch. Dividers may also be used to help transfer measurements to be read when inaccurate or inconvenient to try to read directly from the rule, Figure 2.4.

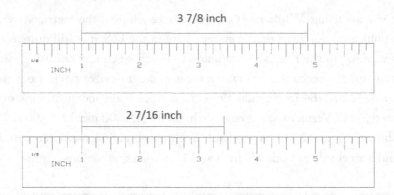

Figure 2.3: Examples of reading a steel rule.

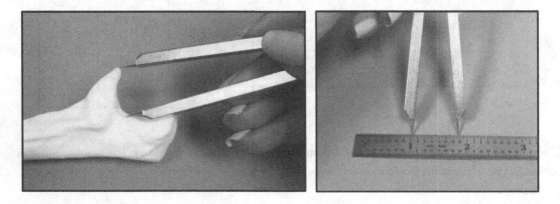

Figure 2.4: Dividers can be used to transfer measurements to rules.

2.2.2   CALIPERS

Calipers are in essence a sliding rule. Although calipers can have an electronic readout, it is useful to understand how to read a dial caliper since it is the caliper style used in machine shops. The parts of the dial caliper are shown below, Figure 2.5.

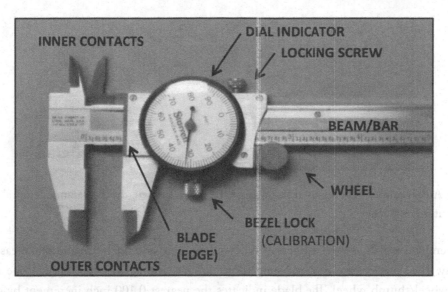

Figure 2.5: The parts of a dial caliper.

The beam of the dial caliper is divided into 0.100 (1/10) inch graduations. In addition, the dial is divided into 100 graduations each equivalent to 0.001 (1/1000) inch. One full revolution of the dial indicates a linear movement of 0.100 (100/1000) inch, which can equivalently be read off the beam as 0.100 inches. The calipers may be used to take both inside and outside measurements, as shown in Figures 2.6 and 2.7, respectively. Using the inner contacts, the inside measurement is made of the ceramic femoral head, while the outside measurement is made of the ceramic femoral head diameter using the outer contacts.

Figure 2.6: A dial caliper is being used to measure the inner diameter of the ceramic femoral head using the inner contacts.

Figure 2.7: A dial caliper is being used to measure the outer diameter of the ceramic femoral head using the outer contacts.

When learning to use the calipers, it is often helpful to add the beam and dial measurements individually. The contacts are brought snug (don't force them) against the object to be measured using the wheel/thumb wheel. The blade indicates the nearest 0.100 inch increment by observing where the blade lies in relation to both the nearest 0.100 inch graduation on the beam as well as the location of the arrow head on the dial indicator. The dial indicates the nearest 0.001 inch increment. For the internal measurement of the femoral head (Figure 2.6) the blade is past the 0.50 inch beam graduation. It is important to be able to view the vertical bar to the right of the visible number on the beam to correctly verify the complete revolution. From the dial the arrow is on the 17th graduation indicating a measurement of 0.017 inch (seventeen thousandths of an inch). Adding the beam and dial yields the final reading or 0.500 + 0.017 = 0.517 inches. For the external measurement of the femoral head (Figure 2.7) the blade is past the 1.100 inch beam graduation. From the dial the arrow is halfway between the 0 (0.000) and 1 (0.001) graduations indicating a measurement of 0.0005 inch (five ten thousandths of an inch). Adding the beam and the dial yields the final reading or 1.100 + 0.0005 = 1.1005 inches. Like the steel rule, the resolution of the dial cannot be approximated beyond the average of the two smallest divisions (in this case ½ (0.000 + 0.001) = 0.0005).

Some additional examples of dial caliper readings are provided in Figure 2.8.

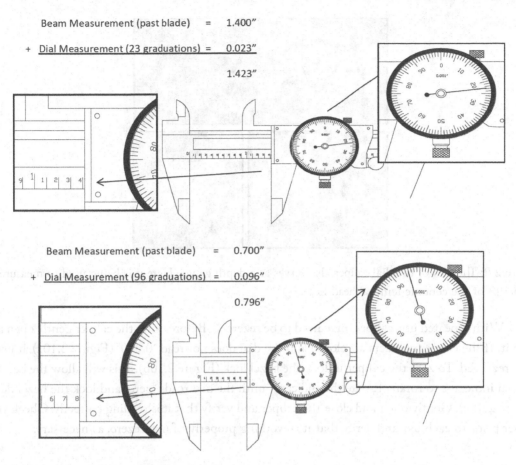

Beam Measurement (past blade)   =   1.400"

+   Dial Measurement (23 graduations)   =   0.023"

1.423"

Beam Measurement (past blade)   =   0.700"

+   Dial Measurement (96 graduations)   =   0.096"

0.796"

Figure 2.8: Examples of dial caliper readings.

In addition to using the dial caliper to acquire measurements with the inside and outside contacts the caliper can also be used to measure depth with the caliper end. As shown in Figure 2.9, the depth of the hole in the ceramic head is measured by moving the thumb wheel (increasing the separation of the outer/inner contacts) until the depth gauge is against the bottom of the hole and the end of the caliper beam is at the outer edge of the implant. The depth measurement is obtained using the exact same approach used to read the outside and inside contact measurements with the beam and dial added for the final measurement.

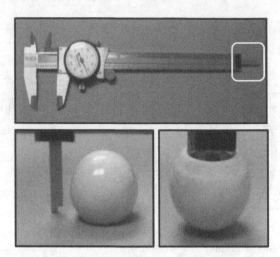

Figure 2.9: The end of the dial caliper also serves as a depth gauge. Here it is being used to measure the depth of the ceramic femoral head hole.

With repeated use, calipers may need to be rezeroed. Before using the caliper gently open and close it. If the caliper contacts are closed and the dial does not read "0.000" (Figure 2.10a), it needs to be rezeroed. To zero the caliper, unlock the bezel lock (Figure 2.10b). This will allow the bezel of the dial indicator to rotate. Rotate the indicator until the dial reads "zero" and lock the bezel down (Figure 2.10c). Gently open and close the caliper and verify that it is reading correctly. Check your caliper prior to each use and verify that it is working properly; if not, rezero, as necessary.

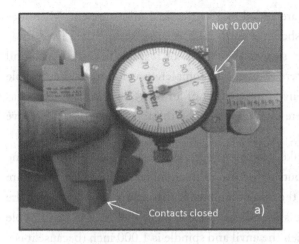

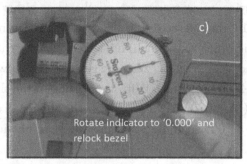

Figure 2.10: (a) If when closed, the caliper dial indicator does not read zero, it needs to be rezeroed. (b) Unlock the bezel lock; (c) turn the indicator dial to read zero and lock down bezel.

2.2.3 MICROMETERS

Unlike the linear/sliding rule used with the caliper, the micrometer utilizes a screw thread. The parts of a micrometer are shown below, Figure 2.11.

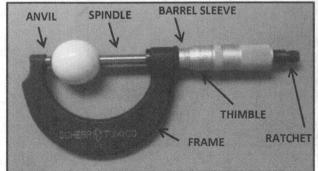

Figure 2.11: The parts of a micrometer.

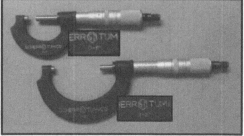

Figure 2.12: Micrometers for obtaining external measurements. (top) A micrometer capable of reading in the range of 0–1 inch; and, (bottom) 1–2 inch are shown.

Graduations on the sleeve and thimble are added to give the final measurement. Micrometers for obtaining external measurements are shown, Figure 2.12. A micrometer has a fixed measurement range. For example, the micrometers shown are capable of reading from 0–1 inch and from 1–2 inch, respectively. For the micrometers shown, one complete revolution of the thimble is equivalent to 0.025 inches (read as twenty-five thousandths of an inch) because the screw has 40 threads per inch (1/40=.025). Four complete revolutions of the thimble, as read on the sleeve represent 0.100 inch.

While standard calipers enable you to take inside and outside measurements with the same device, micrometers are purchased as either outside or inside micrometers. Illustrated in Figure 2.13, the micrometer is being used to measure the outer diameter of the ceramic femoral head. The micrometer (1–2 inch) has a "zero" reading that is equivalent to 1.000 inch. That is, with the thimble and sleeve reading zero, the gap distance between the anvil and spindle is 1.000 inch (because it is a 1–2" micrometer). With the reading of the sleeve at 0.100 and the thimble reading between 0 and 1 ((0.000+0.001)/2 = 0.0005), the outside diameter of the femoral head is read off the micrometer as 1.1005 inch, the same measurement obtained with the calipers.

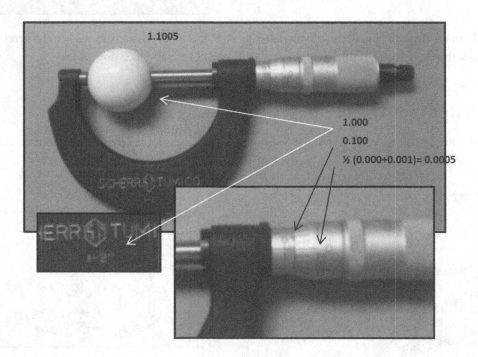

Figure 2.13: A 1–2 inch micrometer is used to measure the outer diameter of the ceramic femoral head.

Some examples of micrometer readings are provided in Figure 2.14.

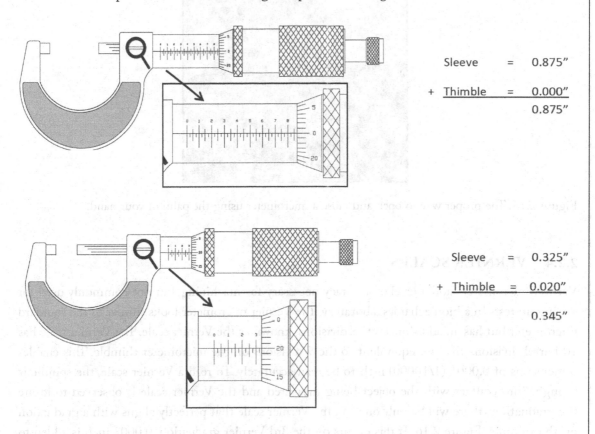

Sleeve	=	0.875"
+ Thimble	=	0.000"
		0.875"

Sleeve	=	0.325"
+ Thimble	=	0.020"
		0.345"

Figure 2.14: Additional examples of micrometer (0–1 inch) readings.

Always use the palm of your hand to open and close the micrometers, Figure 2.15; never spin them. Make sure you understand how to lock and unlock the instruments. These instruments may be used in several ways, for example to verify clearance by locking in at a set dimension. Make sure to unlock the device when you are finished. These are precision instruments that use a ball screw to open and close and they should glide very smoothly. From time to time they may need a little oil, but if properly maintained they should remain accurate for many years.

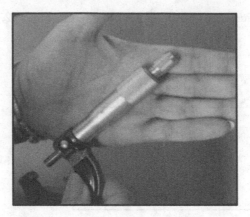

Figure 2.15: The proper way to open and close a micrometer using the palm of your hand.

2.2.4   VERNIER SCALES

Vernier micrometers add a level of accuracy necessary for machining, but not commonly used for setting up tests in a biomechanics laboratory. The Vernier micrometer looks similar to the standard micrometer but has an additional set of divisions, known as the Vernier scale. The Vernier scale has 10 barrel divisions that are equivalent to the 9 divisions on the micrometer thimble. This enables dimensions of 0.0001 (1/10000) inch to be read accurately. To read a Vernier scale, the spindle is brought into contact with the object being measured and the Vernier scale is observed to locate the graduation (there will be only one) on the Vernier scale that perfectly aligns with a graduation on the thimble, Figure 2.16. If this occurs on the 3rd Vernier graduation, 0.0003 inch is added to the object measurement; if this occurs on the 7th Vernier graduation, 0.0007 inch is added to the object measurement. The thimble number is not relevant; the alignment of the divisions is what is matched up to the Vernier scale. This additional measurement is added to the reading from the spindle and thimble to obtain the final reading. Although frequently encountered with respect to the micrometer, the Vernier scale can also be found on other measuring tools including the caliper (Vernier caliper) and the height gauge.

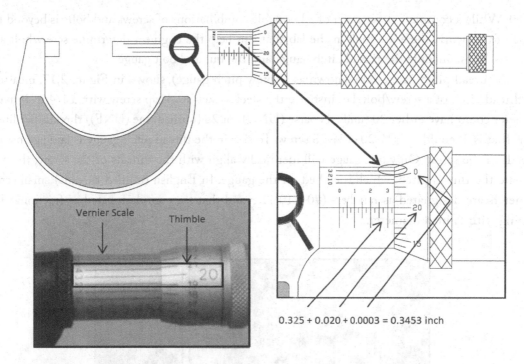

0.325 + 0.020 + 0.0003 = 0.3453 inch

Figure 2.16: A Vernier scale takes the level of measurement precision to 0.0001 inch (one ten thousandth of an inch). The scale can be found on many precision measurement tools, such as on the micrometer shown here. Given where the thimble and Vernier scale line up, 0.0003 inch is added to the dimension.

2.2.5 ADDITIONAL MEASUREMENT EQUIPMENT

There are several additional tools that assist the engineer/machinist in ensuring work is accurate. For machinists these tools may be used in machining setup; for the engineer they can be very useful tools for experimental testing setup. While personal preference will often dictate the types and number of these tools that are found in the laboratory, here are some that we have found to be particularly useful. They include thread pitch gauges, nut and bolt gauges, gauge blocks, feeler gauges, adjustable parallels and digital protractors.

Thread Pitch Gauge and Nut and Bolt Gauges

Screws and bolts are common fasteners that come in a variety of material uses, styles and sizes. They vary by unit system (metric and English) to material use (e.g., for wood and metal). They come with right-handed threads (tightened clockwise) and left-handed threads (tightened counter-clockwise) which may be coarse (for softer metals, e.g., aluminum and brass) or fine (for harder metals, e.g., steels). While a complete description of all possible combinations of screws and bolts is beyond the scope of this work, there are tools in the lab that can help the engineer determine screw/bolt and nut sizes. These include the thread pitch gauge and the nut and bolt gauge.

A thread pitch gauge (also known as a screw pitch gauge), shown in Figure 2.17, measures the thread pitch of a screw/bolt. For instance the steel socket head cap screw with a 1/4 inch major diameter could have either 20 (unified coarse (UNC)) or 28 (unified fine (UNF)) threads per linear inch. That is, it could be a ¼-20 or ¼-28 screw. To size it, the thread pitch gauge is laid against the threads of the screw. Only one gauge will identically align with the threads of the screw; this will indicate the thread pitch, which is etched on the gauge. In English (unified threads) small screw diameters are designated as numbers (#0–#10); beyond that they are designated as fractional, beginning with ¼ inch diameter.

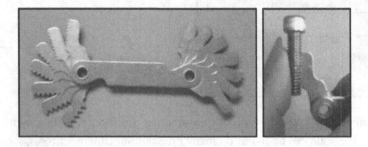

Figure 2.17: Thread pitch gauges may be used to determine the thread pitch of a screw or bolt. As demonstrated, the "20" thread pitch gauge (20 threads per inch (tpi)) fits perfectly into the threads of the ¼-20 socket head cap screw.

Similarly, nut and bolt gauges may be used to size nuts and bolts, or male/female connectors. As shown in Figure 2.18, these gauges size both metric and English (UNC and UNF) nuts and bolts. Here the nut/bolt can be screwed onto/into the correct gauge. The size of the gauge is stamped on its surface. In English the threads are designated by the number of threads per inch (tpi); in metric the threads are designated by thread pitch, or the linear distance between threads (mm).

A socket head cap screw and t-nut for a loading platform mount are shown in Figure 2.19a. The screw, which was previously sized in Figure 2.18 using the thread pitch gauge may also be sized using the nut and bolt gauge. In Figure 2.19b, the socket head cap screw fits into the female end of the ¼-20 gauge, while the t-nut may be sized with the male end of the ¼-20 gauge, Figure 2.19c. Keeping a set of these in the lab comes in handy when sizing existing equipment; determining the size of lost or damaged (e.g., stripped) nuts and bolts for replacement; and when fabricating new fixtures to adapt to existing systems/equipment.

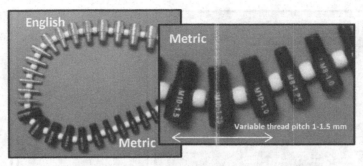

Figure 2.18: Nut and bolt gauges.

Figure 2.19: Nut and bolt gauges are used to size English and metric male and female connectors.

Gauge Blocks and Feeler Gauges

Gauge blocks are precision machined flat (ground) and parallel-faced blocks that come in a variety of sizes and are used as standards for equipment calibration, as well as in setting up machining work and experimental tests. Also known as Swedes, Johansson, or Jo blocks, these blocks may be conveniently purchased in sets, Figure 2.20. In addition to their individual use, owing to surface friction and attraction of the molecular forces the blocks can be "wrung" together to stack them such that a variety of combinations are possible. A light layer of oil should be maintained on the blocks to keep them from corroding, but this is wiped off prior to wringing the blocks. As demonstrated in Figure 2.20, the blocks are wrung together perpendicular to each other (form a cross) and then are twisted relative to each other to finish the stacking process. After use they are separated and stored with a layer of oil.

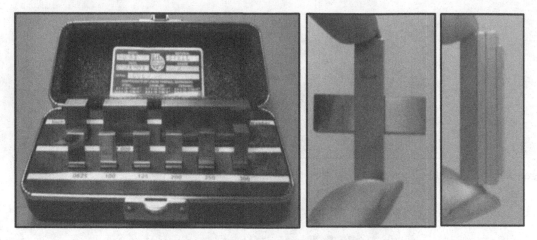

Figure 2.20: Gauge blocks are purchased in kits and can be used individually or wrung together to create dimension combinations that may be used to calibrate equipment and set up mechanical tests.

In Figure 2.21, the blocks are used to check dial caliper accuracy. The .1 inch gauge block (a) and a wrung combination (.1 + .2 = .3) (b) are used to verify the accuracy of the dial caliper, which is plotted in the graph for a variety of standard blocks and wrung combinations.

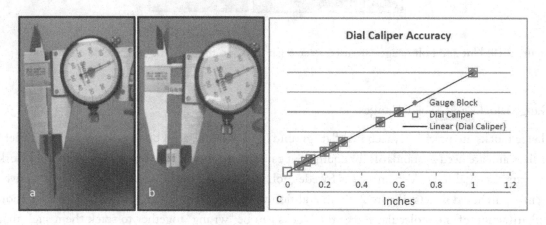

Figure 2.21: Gauge blocks are used to check the accuracy of the dial caliper. (a) A .1 inch block and (b) .1 inch and .2 inch gauge blocks are wrung together to obtain additional standards for calibration. (c) A plot of dial caliper measurement as a function of gauge block dimension verifies the accuracy of the caliper. Once verified as accurate, the caliper is considered calibrated and accurate for its immediate use only. These instruments need to be routinely checked.

In addition to using the blocks as standards in equipment calibration they are convenient in experimental testing setup. For example, they may be used to set the gauge length (gap distance) between two grips/platens in a standard tension or compression test, respectively. As demonstrated in Figure 2.22, a 1 inch gauge block is used to set the gauge length in a tension test prior to inserting the tissue specimen. This standardization step is important to increase testing reliability. For example, end effects due to gripping are present; standardizing the gauge length and holding it constant for all tests eliminates an additional variable that if not accounted for could lead to erroneous data. In Figure 2.23, the gauge block is being used to center the swivel assembly holding the compression platens. This ensures that the contacting faces of the two platens are perfectly parallel to each other. This increases the reliability of the data when testing specimens with parallel faces.

Figure 2.22: Gauge block used to set the gauge length in a tension test.

Figure 2.23: Gauge block used to set distance and parallel in a compression test.

In Figure 2.24a, the 1 inch and the 0.5 inch gauge blocks are wrung together and used to set the distance between lower contacts in a three-point bending test. Although not shown, the distance between the upper contact and either of the lower contacts could also be verified with a 0.75 inch combination. In addition to setting the distance, the gauge blocks may also be used to establish planar alignment between the upper and lower fixtures, Figure 2.24b.

Feeler gauges are similar in concept to the gauge block, but are thinner and generally used to set small distances and measure gaps. Feeler gauges are metal or plastic (useful when dealing with magnetic metals) and the individual gauges (also referred to as "blades") are bound together on a common ring. In this way, they can be used individually or stacked together to obtain a variety of different thicknesses, Figure 2.25.

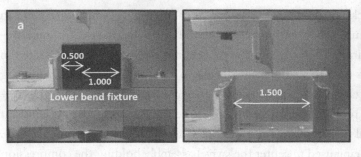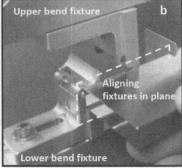

Figure 2.24: (left) Gauge blocks used to set lower contact span in a three-point bend test and (right) align top and bottom fixtures in plane for testing and ensure loading on center.

Figure 2.25: Feeler gauges are a convenient way to set gap distances and to calibrate microactuator movement. Here, three (0,020, 0.015, 0.010) gauges are stacked together and used to verify caliper accuracy.

Adjustable Parallels

Adjustable parallels are similar to gauge blocks, but each parallel consists of two inclined wedges connected via a dovetail joint and a locking screw, Figure 2.26. In their closed state, they look similar to a gauge block but they are not limited in use to discrete measurements. As they are opened up, they maintain parallel edges while their height varies. These tools are ideal for measuring slots, setting distances similar to the gauge blocks and can also be used with calipers and micrometers to measure gap widths in setting up machine tools or laboratory experiments.

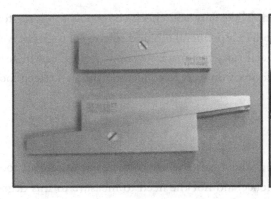

Figure 2.26: Adjustable parallels are useful in measuring gap and slot widths given the inclined plane keeps the faces parallel at all times. (Right) The adjustable parallel is being used to set the contact distance between the lower contacts in a bend test. Here it is being used in conjunction with a gauge block.

Digital Protractors

Digital protractors are useful in verifying/setting a level platform. They may also be used to verify or set angles. As shown in Figure 2.27, a rodent fracture platform attached to the loading machine is not level. Using screw jacks the platform is propped and the digital protractor verifies that the platform is level with respect to the base of the loading machine.

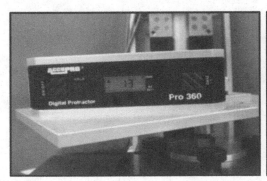

Figure 2.27: Digital protractors can be used to verify level, as in the example above.

2.2.6 HANDLING ISSUES

Precision measuring tools should be wiped down after each use and should be stored with a thin oil film. Individuals have different amounts of oils in their skin and salt in their sweat which can lead to rust on the surface of the tools where handled. Make sure you check your tools for rust, store with a thin film of oil and clean off any rust immediately. Mineral oils and silicon sprays work well. A little steel wool can be used to remove superficial rust, and machine shop "stones" can be used for heavier rust. Camphor cakes can be added to toolbox drawers to collect moisture and keep tools rust-free when not in use.

2.2.7   A PRACTICAL NOTE

One final note on measurement should be made at this time. The use of these measurement tools in our laboratory is based upon practical convenience and not the need to measure to this degree of accuracy and precision at all times. For example, in the case of the gauge blocks, they are ground to an accuracy on the order of millionths of an inch (English unit sets). Here we are not implying that this degree of accuracy and precision is necessary. However, it is very convenient to use these blocks to set experiments knowing that they hold the dimension (to at least the resolution we need ~ 0.01 inch) and that given the parallel faces, they are much easier to work with than using a caliper to make repeated measurements during setup. For example, in the case of setting the distance between lower contacts in a bend test, it is very easy and convenient to place a gauge block between the fixtures and lock them in place using the blocks as a spacer, Figure 2.28. It is also a good way to ensure that even if a student is unable to or errs in reading a caliper he/she is still able to set up a test that is correct and reproducible. In addition, many of these tools serve as standards by which to verify the accuracy of measurement tools (calipers, micrometers) and this is critical to have in a biomechanics laboratory. Moving toward testing at the microscale, concern for increased accuracy and precision will become necessary.

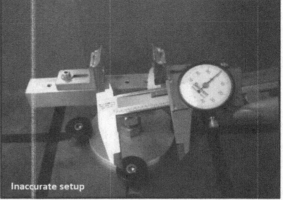
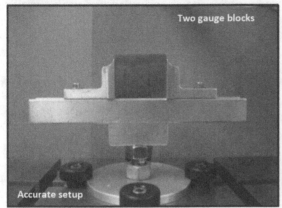

Figure 2.28: While calipers can be used to set a distance, such as shown here for adjusting the lower contact spacing on the bend fixtures, if not held perfectly straight the reading can be erroneous. The gauge blocks (and adjustable parallel) are a simple way to accurately set the distance every time. Many of the tools discussed in this section are used as much for their convenience as for their accuracy.

CHAPTER 3
Design

3.1 MECHANICAL DRAWING

Engineers communicate their ideas to machinists through mechanical drawings. These drawings contain the minimum information necessary and sufficient to correctly create the part. Shown in Figure 3.1 is a loading platform that will be explained in detail later in this text. Here we show the platform with the arm attached. The aluminum arm moves with the linear motion of the machine, and as shown here for three-point bending, enables contact of the upper bend fixture with the bone placed across the lower bend fixture. The linear slide that the arm is mounted to was purchased from a commercial vendor. To use this slide for the linear motion of the loading platform, it was necessary to design and machine an arm that could attach the testing fixtures to the machine.

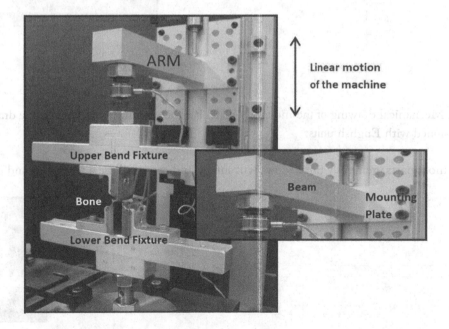

Figure 3.1: Photograph of arm extending from the machine slide via the mounting plate. The plate in turn attaches to the beam to which the fixtures attach.

The arm is assembled from two components: the mounting plate and the beam, Figure 3.2 and 3.3. The mounting plate attaches to the commercial slide. The beam attaches to the mounting plate and in turn holds the fixture(s). The use of the arm enables a variety of fixtures (e.g., bending, tensile, compression) to be attached to the loading machine, as will be demonstrated. The mounting plate and beam were machined separately. For the mounting plate, three views are necessary to completely define the part. Typically three views are provided for parts, but the number is dependent upon the complexity of the part and fewer or more may be needed. If an object is quite complex and involves a lot of angles and hidden surfaces, etc., it may be necessary to provide all six views (front, top, right, bottom, back, left).

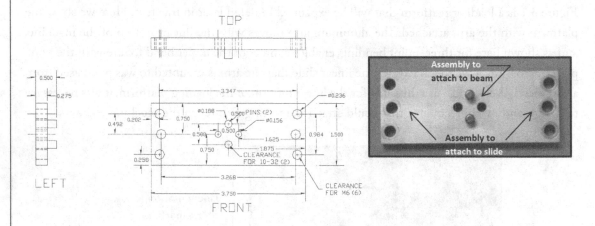

Figure 3.2: Mechanical drawing of the mounting plate. Per the machinist's preference, the drawing was dimensioned with English units.

For most parts, generally three views will suffice and these will be a front, top and side view.

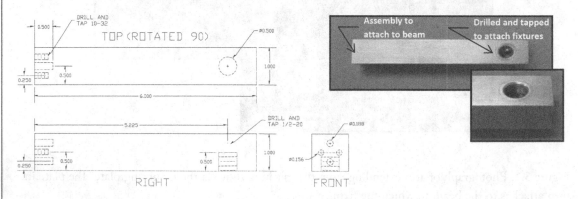

Figure 3.3: Mechanical drawing of the beam. Per the machinist's preference, the drawing was dimensioned in English units.

As previously noted, the information provided in the drawing is the minimal that is both necessary and sufficient to correctly create the workpiece. For example, each dimension should be included only once in the drawing (not in each view) and it should be provided in the most relevant view. A machinist looks at the print and has to visualize the part in three dimensions; redundant dimensions clutter the print and make it difficult to visualize the piece. Most software programs today have a dimension feature that will automatically delete redundant dimensions from a print, but it is ultimately the engineer's job to ensure that only and all the necessary information is provided. In addition, tolerancing is essential. It is important to note when tolerances must be held and when it is not critical. Don't get lazy and decide to hold everything to within one thousandth of an inch. This will significantly increase the time it takes to get the part, significantly increase the price it will cost, anger a machinist beyond belief and leave him/her with the impression that you do not understand your job. This level of tolerancing is necessary only when it is absolutely critical that dimension be "held" (or the exact dimension provided). If the design incorporates mating parts, tolerancing may be critical to ensure that the parts mate properly, for instance fitting over a shaft, held in position with a keyway, or critical to a dimension-based calculation (such as a length dimension). Although it may not seem that critical to the engineer, to the machinist providing tolerances is as important as specifying the material for the part.

While there are many common practices for drafting, many of these are circumvented by the commercial software programs that correctly place dimensions. For example, when measuring linear distances between two holes it is common to reference the dimensions to the center points. Software programs default to these locations. It is important to note if the symmetry is with respect to the center or from an edge and to account for this. For example, the mounting plate shown (Figure 3.4) mates with the beam, and the critical feature is that they mate properly. Furthermore,

Figure 3.4: Mounting plate and beam machined from aluminum. The pins in the plate align the beam on center and machine screws anchor beam to the plate.

for the mounting plate to attach to the linear slide it is necessary that the clearance holes in the mounting plate align with the drilled and tapped hole patterns on the commercial slide. If the parts are slightly off from side to side, it will still mate; if the part is slightly off from center, it will not mate. Thus a few thousandths on the outer length is acceptable, but not on the mating hole patterns. Shown in Figure 3.5a is a sketch of the mounting plate with the desired dimensions. Given that the mounting plate holds the beam and needs to be on center, the center dimensions are critical for the parts to mate and the fixtures to be aligned properly off the beam. As shown in Figure 3.5b, while the same linear dimensions for hole diameter are held, the overall length of the piece is 0.210 inches longer. However, given that the hole pattern allows the plate to mount to the linear slide and the distance between these mounting holes and the hole pattern for beam attachment are the same, this mounting plate will still work as intended. In this case, the only difference is that the overhang on the right side of the plate will be 0.210 inches longer than the left side. While aesthetically this may not be ideal, functionally this has no negative effect. As shown in Figure 3.5c, the mounting plate is the same overall length as the plate in (b). The holes that mount the plate to the slide are not the same as in Figure 3.5a. That is, the linear distance is not the desired 3.346 inches, but 3.556 inches. In addition, the distance between these holes and the hole pattern for beam attachment are not the same. The mounting plate shown in (c) will not work with the beam.

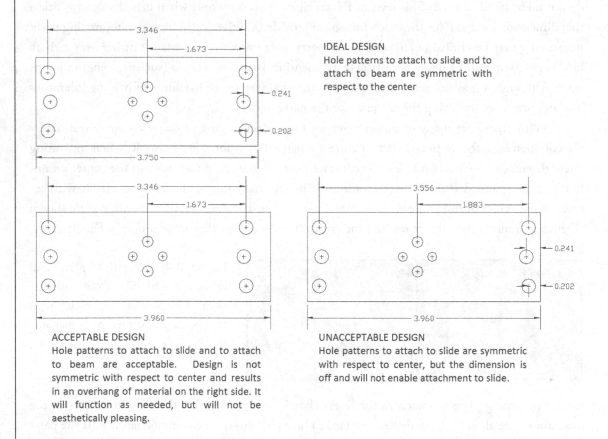

IDEAL DESIGN
Hole patterns to attach to slide and to attach to beam are symmetric with respect to the center

ACCEPTABLE DESIGN
Hole patterns to attach to slide and to attach to beam are acceptable. Design is not symmetric with respect to center and results in an overhang of material on the right side. It will function as needed, but will not be aesthetically pleasing.

UNACCEPTABLE DESIGN
Hole patterns to attach to slide are symmetric with respect to center, but the dimension is off and will not enable attachment to slide.

Figure 3.5: In dimensioning mating pieces, make sure to dimension a piece relative to appropriate symmetry. Improper dimensioning can easily result in pieces that will not align properly.

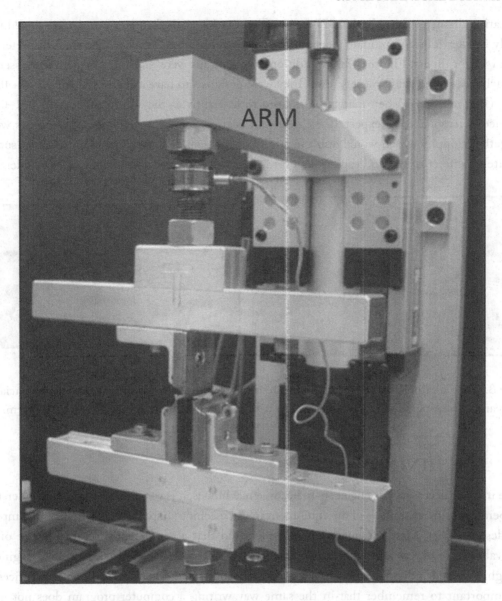

Figure 3.6: The assembled arm holds the upper bend fixture to the linear slide.

Tolerancing is an extremely important concept in design and deals with the degree to which a measurement defined needs to be precise. For instance, in the example shown a rack and pinion torsion fixture was designed and developed in-house to run off of the linear motion of an existing loading machine, Figure 3.7. Figure 3.7c shows the spur gear (pinion) and the bearings mounted in the vertical supports through which a rotating shaft passes. These components were purchased from commercial vendors and selected to match diameters. The spur gear diameter was 0.376 inches and the bearing diameters were both 0.377 inches. The rotating shaft diameter was then designed to be slightly smaller to accommodate the spur gear and bearings, 0.375 inches. The shaft diameter was verified with calipers, Figure 3.7a. In addition, two endcaps were machined to flank the assembly and anchored to the shaft via setscrews. They were designed to have a center hole of 0.376 + 0.002 inches (Figure 3.7b). When mating parts need to fit tightly as they do in this case, 0.001–0.003 tolerance is reasonable to expect. If the gear and bearing centers were larger, components would fit together sloppily which would result in additional motion and error. Other dimensions, such as the outer diameter of the end caps are not critical and do not need to be held to a close tolerance.

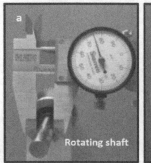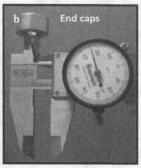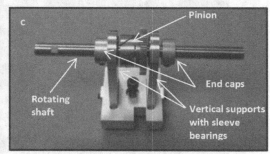

Figure 3.7: The rotating shaft of the torsion system was machined to accommodate the commercial spur gear and bearings. Tolerances were held to within 0.001 inch to minimize slop in the system.

3.2   MACHINING

While it is not necessary for engineers to become machinists, it is valuable to minimally understand the operation and capabilities of the equipment in the machine shop. This will significantly improve your design ability. First, it's important to understand that a competent machinist is one of the most valuable resources an engineer can have at their disposal. Machinists, not unlike engineers, are highly trained individuals that have successfully completed several years of an apprenticeship. It is important to remember that in the same way writing a computer program does not make you a computer programmer, turning a simple cylinder on a lathe or milling a surface finish on a plate does not make you a machinist. Their abilities far exceed those of a novice and remembering that will be important. Do not assume that everyone in the shop is a machinist. Universities often employ shop technicians that do not have the formal training and apprenticeship experience of a machinist. For basic shop work this is often more than acceptable, but an experienced machinist's skills will far surpass those of a technician. Second, with conventional metal machining it is possible to draw an object that cannot be machined. And it is important to understand that just because it can be drawn, doesn't mean it can be made. By understanding the basics of machining this will become more obvious with experience and practice and you will understand how to design around these limitations. Furthermore, what cannot be machined in a conventional shop may be able to be fabricated with an electrodischarge machine or contact printing techniques (stereolithography with metal plating). In rapid prototype systems, plastics are dropped and hardened in thin layers building up the device from a 2D planform to the 3D object. These systems enable intricate geometries (particularly internal geometries) to be created that may not be able to be fabricated by conventional machining. But for a variety of reasons beyond the scope of this text, they may not be suitable for basic biomechanics research. Here we will focus on conventional machining that has traditionally been used to machine metal fixtures for biomechanics studies.

3.2.1 MACHINE SHOP SAFETY

Before even stepping foot into a machine shop it is important to review safety procedures. OSHA, which stands for the Occupational Safety and Health Administration, is the federal agency established in 1970 that is responsible for safety guidelines practiced in the machine shop. Machine parts are moving at fast speeds and loose fitting clothing is easily caught. Do not wear jackets and clothing with drawstrings; remove all jewelry, hats and head scarves. Long hair should be pulled up and fastened securely and close to the head. Toes should be covered, preferably with steel toed shoes, and eye protection should be worn at all times. Never operate any piece of equipment sitting down and make sure that benchtop machines are firmly bolted to a stationary bench. Safety cannot be stressed enough as it takes only a second of lost concentration to lose a finger or a hand. Do not assume if an injury (such as a puncture) does not bleed it is not serious. Some of the most dangerous wounds are wounds that do not bleed; allowing bacteria to build up in the body becoming serious infections. ALWAYS BE AWARE OF YOUR SURROUNDINGS. Always think through the machining operation before you conduct the work. For each piece of equipment, make sure you understand what part of the machine is cutting and locate all emergency switches. Watch overhead lighting when handling long pieces of stock material; there is a serious risk of electrocution, particularly in smaller shops where ceiling heights are low or suspended, uncovered fluorescent lights are used. Make sure there is a fire extinguisher in the room and you are aware of how to use it. If you are on any medications, make sure that it is safe for you to be working in the shop and let the shop attendant know you are on these medications. Some medications cause drowsiness that can result in injury. Avoid conversations and other distractions in the shop. If you do not know what you are doing in a machine shop this is not the place to turn the handles on the machines to see what happens. Most importantly, NO ONE, regardless of skill level, should ever work alone in a machine shop (Figure 3.8) and no one should be allowed to work in a machine shop until they have successfully completed a training course on equipment operation conducted by a qualified machinist, or trained shop technician.

Figure 3.8: Safety should be the number one priority in any machining operation.

3.2.2 STOCK MATERIALS

When designing fixtures or loading platforms, it may be necessary to order stock materials. This could be for example a 12 inch * 12 inch * 1 inch plate that is going to be cut into smaller pieces for fixtures or it may be a cylindrical stock piece. Particularly with plate stock it is important to realize that it may not be truly flat or straight and it may be necessary to make the material flat and straight. The surface can be skimmed in a mill to level. It is not uncommon for stock material to range over the material by at least 0.020 inch. Only skin the work piece if it is absolutely necessary. If tolerancing is not critical to the fixture this variance is probably acceptable. If tight tolerances are necessary, it may be necessary to add this step. At the very least measure the stock pieces to determine the extent to which the dimensions vary across and along the material. As shown in Figure 3.9, a dial indicator inserted into the chuck on a milling machine can be swung around the surface of the plate to determine if the plate is flat and level. The indicator's tip is slightly depressed on the plate surface and the indicator is traced across the surface of the work piece. If the tip loses contact with the plate at any point, the indicator is reset and the tip is depressed deeper into the plate. This is repeated until the tip just barely remains in contact with the surface at all times. The dial is then zeroed and the indicator is then swung again and the amount of variation in the surface is noted. Because the surface will have regions that are higher and lower than the tip level, the gauge will fluctuate in both directions around zero. The indicator dial is read in the same way the dial of the dial caliper is read. For this operation, the plate is held on the milling machine via a vise. In addition, the plate is set upon parallels to keep the plate above the vise surface and the parallels ensure that the plate rests equally on both parallels at a uniform height with respect to the vise. If parallel is critical, the top surface is skimmed then flipped and the bottom surface is skimmed relative to the top surface.

 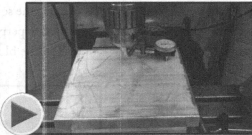

Figure 3.9: A stock plate is placed in the mill and an indicator is swung to quantify the variation in plate thickness.

Familiarize yourself with industry catalogs that provide great insight into purchasing stock materials and machine tool equipment. McMaster Carr catalogs are books unto themselves that offer explanations for example on what particular fasteners (screws/bolts/nuts/pins…) are used for and how they are designated by size. If you're not sure if you need a socket head cap screw, lag screw or shoulder bolt, this book will help you determine this. It is also worth noting that many machinists still use the English unit system even though engineers are taught to work with the metric system. Therefore, to avoid confusion, always make sure if you are talking about drilling and tapping a hole for a "6" screw that the machinist is not referring to a 6-32 (English size 6 (0.138-inch diameter), 32 threads per inch) if you want an M6 (metric size M6-0.75; metric 6 (6 mm diameter) with 0.75 mm thread pitch).

3.2.3   DESIGN LAYOUT

In the same way that an engineer will lay out a drawing on paper (or a computer screen) a machinist will lay out the location of features to be machined on the work piece. This can be the location of a center hole to be drilled and tapped in a plate to accept a threaded connection or it could be a line to indicate where to cut the material to size. For instance, if a stock aluminum plate that is 12 inch * 12 inch * 1 inch is cut into three smaller blocks that are each 3 inch * 12 inch * 1 inch, a band saw can be used to cut the plate. Layout dye is used to paint the surface where the cuts are to be made and a combination square and scribe is a useful way to mark the material. In lieu of layout dye, a permanent marker may also be used to color the surface before scribing a line. A gentle, dry cleaning pad (Scotch-brite™) is useful at removing any ink from the final part.

The combination square and its parts are shown labeled in Figure 3.10a. The set shown is based upon a 6 inch blade. Longer blades are available, as are beveled protractors for angular measurements. In Figure 3.10b the square head is being used to score a line on the aluminum block indicating where a cut is to be made. To do this the square head rests flush against the left side edge of the block while the blade is used to score a line perpendicular to the left edge. To easily visualize the score line, permanent marker was drawn on the block prior to running the scribe along the blade.

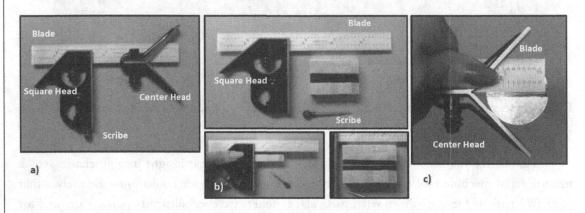

Figure 3.10: (a) A combination square and its labeled components. (b) The square head is used to score a straight line across the work piece. (c) The center head is used to locate and mark center of a round work piece.

While there are several tools that help with layout work, the combination square is particularly useful. It can be used to draw lines as shown above (Figure 3.10b) and it can also be used to draw longer lines than the blade can accommodate. To do this, the square head is placed along the left edge of a work piece. The blade is set such that its right edge sets the desired width. The scribe is placed on the right edge of the blade and held against the blade while a sliding motion is used to move the tool along the length of the piece and score the line to the required length. The center head may be used with round cross sections to quickly locate center. For instance, as shown in Figure 3.10c, with the work piece contacting both sides of the center head, the blade is coincident with a line running through the center.

3.2.4   THE EQUIPMENT

The intent of this section is not to explain how to use the machines, but to demonstrate what they can do to begin to demonstrate how simple work pieces are machined. Furthermore, in the same way that a part can be drawn in many ways, there are different ways to machine a part. For example, a hole may be drilled with a drill press, milling machine or lathe. While we acknowledge that there are several ways to machine a work piece, we will try to provide a few examples to illustrate common uses of the equipment.

The Band Saw

Band saws are used to take rough cuts of a material. With a band saw, the blade of the saw moves against the stationary piece of stock material that is being cut. Band saws may be either horizontal or vertical, Figure 3.11. Horizontal band saws are typically referred to as "cutoff" saws, again indicating their use in rough cutting materials. Before cutting with a band saw, make sure the safety guard is placed just above the thickness of the work piece to be cut. Depending upon the size of the cuts needed, hacksaws may also be used. These manual saws are great for cutting smaller rods and softer materials such as aluminum. If cutting with a hacksaw, make sure to mount the material safely in a vise and never let someone else hold the material while it is being cut. Furthermore, always look at your blade to determine the direction of the teeth. If the hacksaw does not cut on the forward stroke, the blade is in backwards.

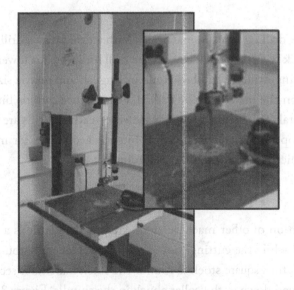

Figure 3.11: Typical vertical band saw for rough cutting stock materials.

The more precise the cuts made on the saws, the less mill work needed to clean up the cuts. Saws also have a thickness to them, so if 3, metal strips that are precisely dimensioned to 4 inch * 12 inch * 1 inch are needed and the plate is precisely 12 inch * 12 inch * 1 inch, larger stock plate is required. The thickness of the blade produces a loss of material as it cuts. This is known as the "kerf" of the saw and includes not only the thickness of the blade, but the wobble in the cut which can result in a significant loss of material. If rough cutting a plate, make sure to be on the outside of the scribed line to ensure that the desired dimension even with the kerf can be obtained. The rough cut is cleaned up in the mill.

When operating a saw, never push the material through the blade. Gently guide the material to be cut, letting the blade do the work. Do not force it; this is one of the easiest ways to break a blade which can be very dangerous. Also check that the blade is sharp or it will not cut properly. Only move in a forward direction slowly guiding the work into the blade; do not reverse directions and go backward—the teeth are only on one side of the blade and it will only cut in one direction. If the blade is broke, or missing or needs to be replaced and you have never changed one, do not attempt it on your own. Seek out help from someone with experience. Sometimes the blades are purchased as long strips. On the front of the saw is a welder that will let you cut the blade to the desired length and weld it together then anneal it for a strong bond. Again, seek assistance if you have never done this before and also ask for help in disposing of the original blade. Never leave blades lying around the shop.

The Drill Press

Drill presses are used for drilling holes in materials. With a drill press, a drill bit is secured between the center jaws of a chuck with the chuck key. As the drill bit spins it is lowered onto the stationary stock material that is being drilled. Drill bits come in English and metric sizes and are of different hardness depending upon the material to be drilled. For instance carbide bits are particularly hard and ideal for cutting metals that are heat treated. However, carbide drills are brittle and easily shattered if not handled properly. Standard steel drill bits are acceptable for most fixture machining required in a biomechanics lab, such as drilling aluminum.

The Lathe

In contrast to the operation of other machine shop equipment, a lathe is a machine in which the stock material is turned while the cutting tool is stationary. Generally round stock is turned in a lathe, but it is possible to turn square stock to round. The stock material is secured in the lathe using the headstock and clamped down in the collet chuck in the spindle, Figure 3.13. The carriage holds the tool post that accommodates the cutting tool and the cross slide and compound rest that feed the tool into the work piece. When turning on a lathe the amount of material extruding from the headstock should only be slightly longer than the desired length of the finished part. Leave enough room to allow for cutting the finished part off of the stock, but not too much that it is not securely held in the lathe. The shorter the length extruding from the headstock the "truer" or more round the material will spin in the lathe and the more accurate the machining.

Figure 3.12: Typical metalworking lathe most commonly used for machining round stock, drilling holes and cutting threads.

Figure 3.13: Three jaw chucks, as opposed to four are generally self-centering; four jaw chucks hold square stock.

Indicators are gauges that can be placed on the lathe (via a magnetic base) and can be positioned to touch the outer edge of the stock material. As the lathe turns (manually, with motor in neutral) the extent to which the part is out of round may be checked. The indicator may be slid on the lathe along the length of the part to verify the extent to which the part is running "true." While this may seem to be beyond what you as an engineer need to know, understanding these procedures will help you to appreciate the difficulty a machinist has in holding a round part of reasonable length (~ 6 in) to a high tolerance (0.0005) and hopefully it will help you avoid designing to these tolerances unless absolutely necessary.

The lathe is the ideal machine to turn down round stock, drill through holes, drill and tap interior holes in round stock material and cut external threads. Tap charts are found in every machine shop, machinist handbook and trade catalog and indicate the hole size that should be drilled to then tap a hole. The tap is a cutting tool that is used to cut threads into the material. Using a metal drill bit (the same as used in a drill press), the drill bit is inserted in the three jaw chuck of the lathe riding on the ways and spun to cut the hole, the material is left in the machine and the tap is used to cut the threads.

The Milling Machine

The mill is a machine in which the stock material is mounted stationary to the table while the cutting tool (end mill) is turned into the work piece to cut, Figure 3.14. While there are a lot of uses for a milling machine, the mill is often utilized with machining operations on a metal plate. As noted previously, stock material from a supplier may not be perfectly flat. If the final design requires the plate to be uniform, the surface may be cut down on the milling machine, as shown below, Figure 3.15. Here the collet is used to hold the fly cutter that holds the cutting tool. As the fly cutter turns and the table is moved from left to right to move the work piece into the cutting tool, the surface is skimmed level with the passing of the tool. Note that the work piece is placed such that the cutting tools are driven off the edge of the work piece and the entire surface is faced (shiny portion). In addition, to ensure that the plate is skimmed level, prior to resurfacing the plate is clamped in the vise on a pair of parallels.

Another common procedure completed on a mill is to cut into a work piece. This can be everything from a narrow recess to a full thickness slot. As shown in Figure 3.16, an end mill is being used to cut a recess into the plate. Given the amount of material to be removed, the depth of the cut may be taken in one pass or may require multiple passes. If the depth is not taken in a single pass, the milling machine table is raised to bring the work piece into the cutting tool.

For lathes, drill presses and mills there is a simple formula to calculate machine revolutions per minute (RPM). The RPM is a function of the diameter of the cutting tool or work piece (D) and the cutting speed (CS) which is a function of the material used. That is, RPM = (CS *4)/D. D for drill presses and mills represents the diameter of the drill bit or end mill. For the lathe in which the material turns, the diameter of the stock round material is used.

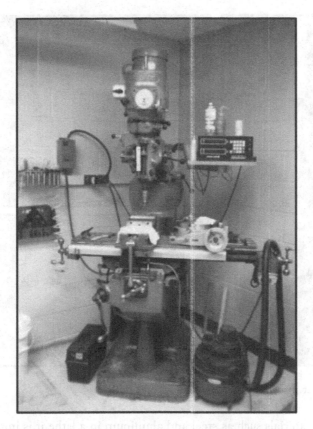

Figure 3.14: Typical milling machine generally used for working with flat plate.

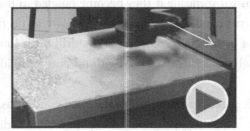

Figure 3.15: A fly cutter inserted into the mill via a collet holds the cutting tool. As the spindle turns the fly cutter the cutting tool sweeps the surface of the plate. The surface can be skinned to level or cut to a specific thickness in this procedure.

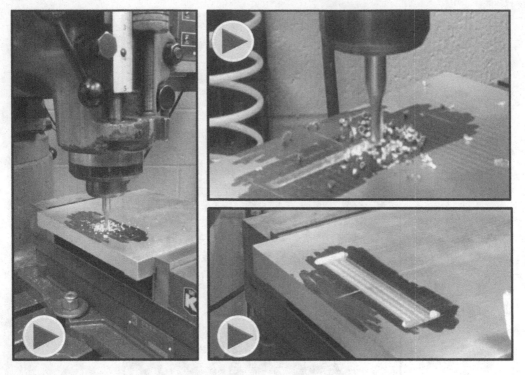

Figure 3.16: A slot is cut into the flat plate with an end mill.

When cutting materials such as steel and aluminum in a lathe it is important to "break" the chips that are being created with the removal of the material. These chips can get quite large and become very dangerous for someone if they become tangled in them. It's always smart to break the chips by moving the cutting tools back and forth slowly as needed. The same is true for drilling holes with a drill press or milling machine. As the drill bit is lowered into the material, raise and lower the drill bit slowly as it cuts the hole (with oil/cutting fluid) to relieve the stress on the tool and break the chips.

In general, work pieces are cut under well-oiled conditions. This can be very messy in small shops and can make maintaining a clean working environment difficult. Sometimes machine rates can be slowed to reduce the need for cutting fluids. If you are working with these machines, make sure you understand the extent to which oil is used and the extent to which operating conditions are dictated by these fluids.

3.2.5   THREADING

Taps and dies are used to cut internal and external threads. Taps cut the internal threads into the female mating piece in a process referred to as tapping; dies cut the external threads into the male mating piece in a process referred to as threading. Taps vary with respect to the shape (tapered or flat (bottom)) and the material from which they are ground and the number of flutes (or threaded regions) along the tap. The flutes cut the threads, while the non-fluted regions provide the space needed for the chips to escape.

Taps may be manually driven into a part, or driven into the piece under the machine power of a lathe or milling machine (power tapping). Regardless of the manner in which the tap is driven, it is critical that the tap be used with cutting oil, that it be run perfectly straight and that care be taken to slowly drive the tap in and out of the work piece to break the chips. It is also important to be aware of the resistance of the material against the tap as they can be very brittle and are easily broken. Shown in Figure 3.17 is a convenient way to tap a hole. Although this tap is being manually driven (no electric power) into the work piece, the milling machine is being used to keep the tap running straight. That is, the chuck is being used to hold a spring loaded center that is applying compression to the tap. Under cutting oil the tap is slowly turned clockwise (~ 1 turn) and then backed out (~ ¼ turn) to cut the threads. As the tap is threaded into the work piece, the drill chuck is continually lowered to maintain compression on the tap. Once the tap is through, the spring-loaded center is removed and the tap is slowly backed out of the work piece. To finish the thread a chamfer tool is placed in the chuck and used to deburr and add a professional edge to the hole. For the best result a simple approach is to jog (turn on and off) the machine and while off, but still spinning, drive the chamfer into the work piece. With a little practice the jog will provide enough power to chamfer the face. If the power were left on the chamfer tool would be pulled into the work piece, which should be avoided.

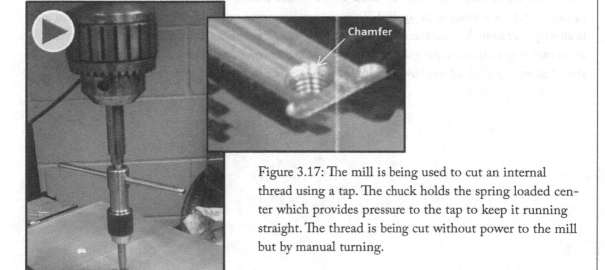

Figure 3.17: The mill is being used to cut an internal thread using a tap. The chuck holds the spring loaded center which provides pressure to the tap to keep it running straight. The thread is being cut without power to the mill but by manual turning.

3.2.6 FIXTURE FABRICATION EXAMPLE

The series of photographs that follow depicts the machining process used to fabricate a set of serrated friction clamps for holding elastomeric materials. In addition, a cross brace was added to keep the fixtures at a fixed distance with the goal of growing bone cells on the elastomeric substrate. The finished fixture, Figure 3.18, was machined from Plexiglas and polycarbonate and consisted of mating top and bottom serrated Plexiglas fixtures connected via socket head cap screws (#4–40). The polycarbonate brace was slotted to accommodate the elastomeric substrates in a variety of distracted states and held to the fixtures via socket head cap screws (#10–32).

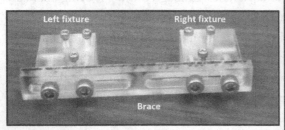 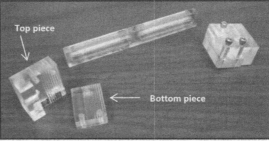

Figure 3.18: The following series of figures depict the machining of the loading fixtures shown here. The fixtures consist of two serrated friction clamps (a left and a right) and a cross brace. An intended use of these fixtures is in the loading of elastomeric substrates. By growing cells on the substrates, they can be subjected to mechanical loading and the cellular activity can be quantified. The substrates are a silicone base and have a tendency to "stick" to the culture dishes; hence manipulation of the substrates causes stretch or loading to be applied. To avoid any unwanted pre-load, the cross brace is applied and is an easy way to make sure the membranes are not loaded in culture. In addition, the brace serves as an easy way to transfer the substrates to the loading machine and also to keep the cells in the stretched state post-load to allow for staining and imaging in the distracted state.

Plexiglas work piece is placed in mill in vise and elevated on parallels.

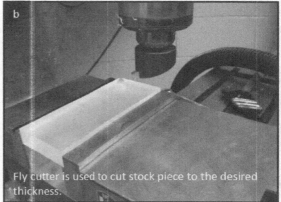

Fly cutter is used to cut stock piece to the desired thickness.

Edge is cleaned up with an end mill. Marker is used to verify that the entire surface is faced.

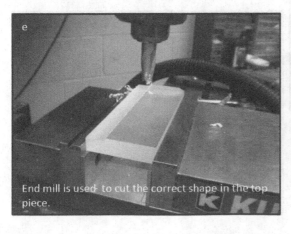

End mill is used to cut the correct shape in the top piece.

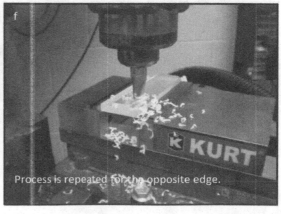

Process is repeated for the opposite edge.

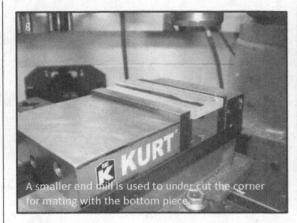

A smaller end mill is used to under cut the corner for mating with the bottom piece.

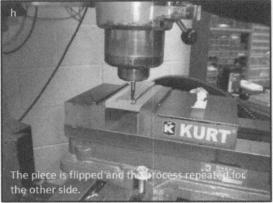

The piece is flipped and the process repeated for the other side.

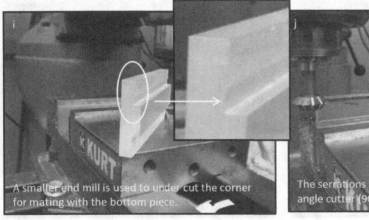

A smaller end mill is used to under cut the corner for mating with the bottom piece.

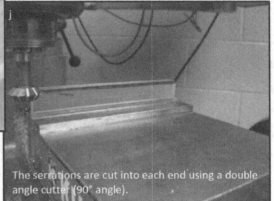

The serrations are cut into each end using a double angle cutter (90° angle).

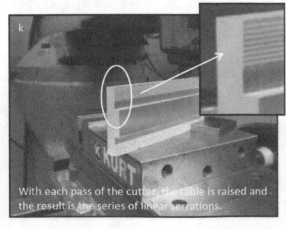

With each pass of the cutter, the table is raised and the result is the series of linear serrations.

The individual fixtures are rough cut to size on the vertical band saw.

The individual fixtures are rough cut to size on the vertical band saw.

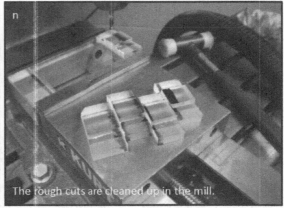

The rough cuts are cleaned up in the mill.

The work pieces are returned to the band saw and cut in half.

The work pieces are returned to the band saw and cut in half.

The work pieces are lined up in the vise and cleaned up to final size in the mill.

The work pieces are lined up in the vise and cleaned up to final size in the mill.

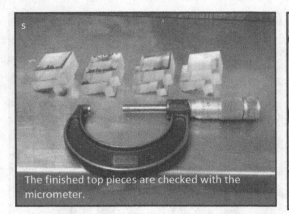

The finished top pieces are checked with the micrometer.

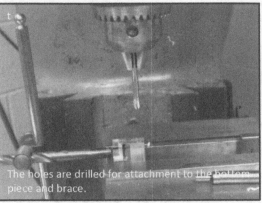

The holes are drilled for attachment to the bottom piece and brace.

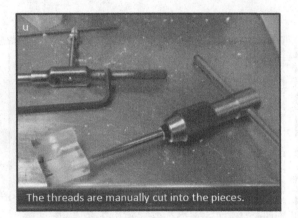

The threads are manually cut into the pieces.

Finished top piece

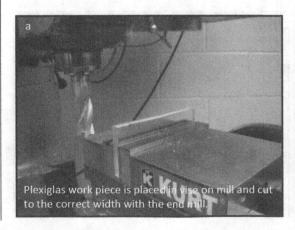

Plexiglas work piece is placed in vise on mill and cut to the correct width with the end mill.

Plexiglas work piece is placed in vise on mill and cut to the correct width with the end mill.

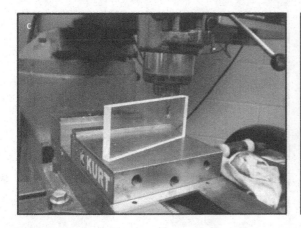

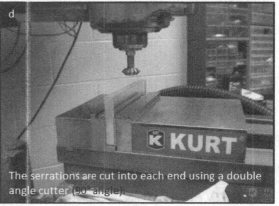

The serrations are cut into each end using a double angle cutter (90° angle).

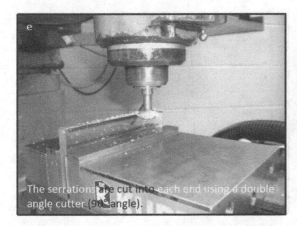

The serrations are cut into each end using a double angle cutter (90° angle).

The serrations are cut into each end using a double angle cutter (90° angle).

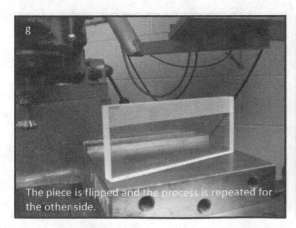

The piece is flipped and the process is repeated for the other side.

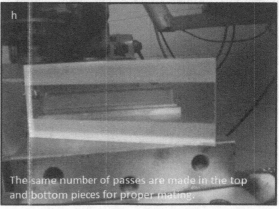

The same number of passes are made in the top and bottom pieces for proper mating.

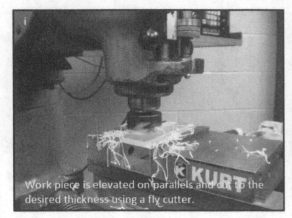

Work piece is elevated on parallels and cut to the desired thickness using a fly cutter.

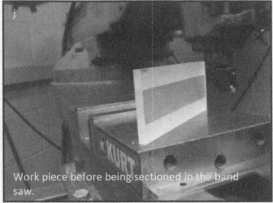

Work piece before being sectioned in the band saw.

Rough cuts area cleaned up in the mill.

Edge is cleaned up with an end mill.

The work pieces are returned to the band saw and cut in half.

q

Finished bottom piece with three drilled and tapped holes for connection to the top piece.

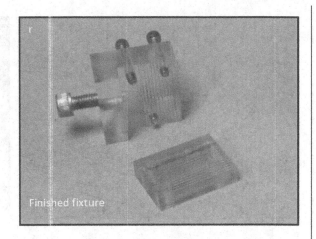

r

Finished fixture

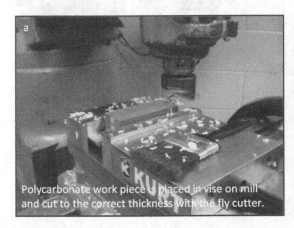

a

Polycarbonate work piece is placed in vise on mill and cut to the correct thickness with the fly cutter.

b

The work pieces will be halved. A marker and scribe are used to layout the cuts to be made.

c

The work pieces are rough cut in half using the band saw.

d

Two at a time, the work pieces are elevated on parallels in the vise and cleaned up with the fly cutter.

Two at a time, the work pieces are elevated on parallels in the vise and cleaned up with the fly cutter.

Two at a time, the work pieces are elevated on parallels in the vise and cleaned up with the fly cutter.

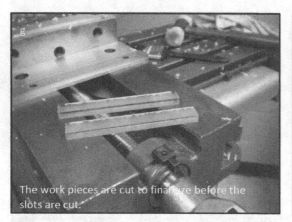

The work pieces are cut to final size before the slots are cut.

The work pieces are cut to final size before the slots are cut.

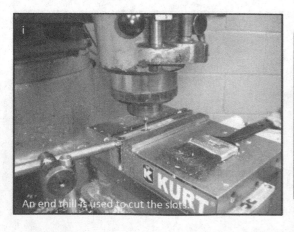

An end mill is used to cut the slots.

Screw clearance is verified in the finished brace.

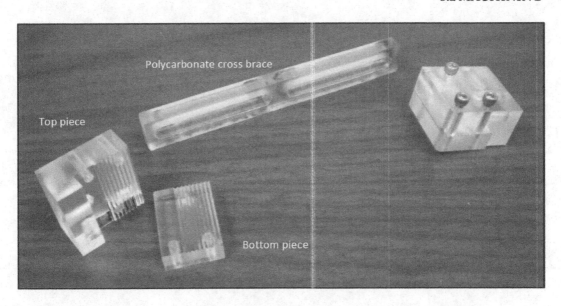

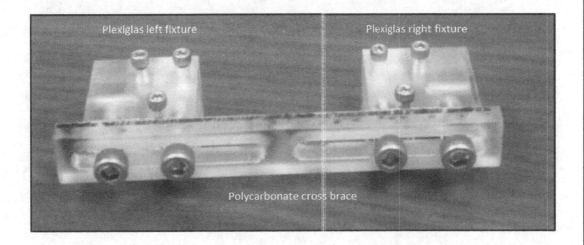

CHAPTER 4
Testing Machine Design and Fabrication

4.1   MECHANICAL TESTING

Mechanical testing machines grew out of a need to determine the load-bearing ability of objects and the materials from which they are fabricated. From a theoretical standpoint, one can use Newton's Laws and fundamentals from physics and material science to approximate behavior, but experimental validation is critical. What is needed is a method to apply a load and measure the specimen deformation as it relates to, for example, length and/or shape changes.

4.1.1   FORCE MEASUREMENT

In the most basic form, a measurement can be binary—such as a simple "yes" or "no" with respect to the ability of the load to break the specimen. The qualitative nature of this measurement is not very useful by itself and quantitative data is needed. Quantitative data such as the magnitude of the load that broke the object and the amount of deformation prior to failure are very useful pieces of information. Systems to obtain quantitative data can be very simple. A bucket suspended from the object can be used to hold weights added individually until the object breaks. The finer the increments (the smaller the weights), the more precise will be the data. Studies have used everything from large weights and washers to BB and buckshot pellets. The discrete increments result in studies that are not as well-controlled as they could be, but do offer a vast improvement over a binary (break/no break) test. To improve upon the controllability of the load, load can be applied in a variety of ways with some of the more common methods including electric (motors), pneumatic and hydraulic means. Analog signals can be used to track both the change in the load as well as the change in displacement of the device driving the load. Electronic systems incorporate piezoelectric (crystal), reluctance/inductance (air gaps) and resistance (load cells).

Load Cells

The most common analog force measurement device in biomechanical testing is the load cell. Testing machines are augmented with these transducers to record the load during a test. These highly accurate sensors are critical to obtaining reliable, quantitative loading data during the test. Common load cells for biomechanical testing systems utilize strain gauge technology.

Strain gauge-based load cells have a sensing junction/element that when loaded responds to the load by producing an electrical output. The circuitry to convert the response to a quantifiable electrical signal is a configuration known as the Wheatstone bridge. The bridge, commonly having four arms, is balanced and temperature compensated. One of the arms is replaced with the load cell strain gauge and an excitation voltage is applied. When the load is applied to the load cell and the sensing element is strained, the resultant change in resistance is read. Knowing the calibration information for the load cell (obtained from the manufacturer at the time of purchase), the voltage change is converted to the applied load. The latter conversion may be handled automatically within the software of many commercial systems or may be manually applied at the time of data analysis for simple machines often developed in-house.

Styles of load cells vary but utilize similar strain gauge technologies. Strain gauge technologies are commonly used in canister, bending, shear and torsion load cells. In these systems the concepts are similar in that bonded strain gauges are fully bonded to thin structural elements that strain upon loading. This is illustrated in Figure 4.1 for a typical S-load cell (bending) and shear load cell. The S-load cell shown offers a relatively easy way to obtain high strain (larger deflections) for comparably low loads. The shear load cells are precision machined with a symmetric cutout on each side of the beam leaving a cross section that is similar in appearance to an I-beam. And, as in an I-beam the load (shear load) is carried by the web (central vertical support). Strain gauges bonded to each face of the web enable the measurement of pure shear. Any bending stress is carried by the flanges (horizontal supports) of the beam and considered negligible.

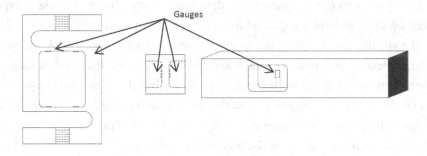

Figure 4.1: Schematics of bending (left) and shear (right) load cells.

Canister load cells are one of the more common load cells found in biomechanics laboratories, Figure 4.2. Canister load cells are relatively robust and have the added benefit that they can easily accommodate both tension and compression. Load cells, particularly for small loading ranges, require very thin backing and are not particularly practical below 50 grams. In addition, they can be purchased in a variety of loading ranges with a relatively low profile, such that a 50 gram canister occupies approximately the same space as a 445.0 N canister. This aids in design when developing testing fixtures to accommodate a variety of testing needs, such as varying animal species or working within a set space for a given loading platform.

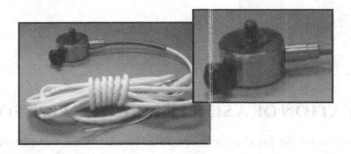

Figure 4.2: Canister load cells come in a variety of loading ranges, and the relatively small profile is ideal for biomechanics work.

4.1.2   DISPLACEMENT MEASUREMENT

Devices are also available to measure displacement. Absolute changes in length can be measured most simply with a ruler. More precise measurements can be obtained with a variety of tools such as calipers and micrometers. Given the practical use of these devices in the biomechanics laboratory, these were discussed at some length previously.

Analog devices are also available and commonly include linear variable differential transformers (LVDTs) (Figure 4.3) and extensometers. Devices such as extensometers and fully bonded strain gauges directly measure change via a physical attachment to the object being tested. As long as they do not interfere with the measurement, they can be quite useful. However, their usefulness is limited. For example, a strain gauge (in the form of a strain rosette), may at most provide a three-dimensional point strain, and to map out the strain profile several strain gauges are necessary. Extensometers work by physically attaching to the specimen being loaded. For relatively rigid materials such as metals, they can be quite useful. However, for more elastic materials (ligaments and tendons) if they hang from the specimen body they can interfere with the measurement being taken and possibly cause premature failure. For this reason, if using an extensometer they should be mounted to a base frame rather than suspended from the specimen. Expensive systems using laser technologies are also available. However, for basic biomechanics work simple non-contact methods of elongation monitoring exist (video dimensional analysis) and can be very useful for determining strain changes. In addition, with the capability of today's digital systems, a standard video camera has the necessary resolution to measure changes in surface contrast markers. For rotational motion, rotary variable differential transformers (RVDTs) are available.

Figure 4.3: Linear displacement sensor.

4.2   FABRICATION OF A SIMPLE LOADING PLATFORM

A mechanical testing machine for the purpose of this book will be thought of as a system that can minimally apply axial movement. Analog measuring devices can then be added to record the load and displacement during movement. The motion of the machine can be driven by a variety of power sources including electromechanical, servohydraulic and pneumatic. Systems are chosen based largely upon given needs. For example servohydraulic systems have large loading capabilities and the ability to accommodate complex loading waveforms. On a negative side, they can be costly, messy to maintain and dangerous. The pistons are under tremendous pressure. Particularly when testing to failure under load control, catastrophic failure can result in loading instability. This instability occurs because as failure occurs a large load drops to "0" within fractions of a second. These systems need tight feedback control to minimize possible injury and machine damage. By comparison, mechanical systems such as screw-driven systems are safer to operate, cleaner and less costly to maintain. These systems work well for slow movement and strain rates. However, given the mass typically found in the crosshead these systems are not ideal for fatigue. Commercial systems for fatigue are available with loading frequency capabilities on the order of 100 Hz.

Significant improvements to testing machines in recent years have increased their regularity and practicality in biomechanical testing. Advances have enabled smaller loads for applications such as small rodent tissue testing, scaffold testing and cellular stimulation. Low scale (high cost) specialty devices on the order of picoNewton loads are also available as are planar biaxial systems. Environmental chambers and submersible load cells enable systems to be turned into platforms that allow tissues to be dynamically tested in hydrated environments such as in developing a bioreactor for stimulating cell-seeded scaffolds under long-term cyclic loading for regenerative medicine applications. The latter will be briefly discussed in Chapter 8 (Mechanobiology).

4.2.1   MECHANICAL TESTING PLATFORMS

Commercial testing machines for biomechanics are generally servohydraulic, pneumatic or electromechanical. Each system has advantages. For example, servohydraulic systems offer a quick response time to the controller signal. This response time results in these systems being ideal for extreme loading rates, either very high or very low. The servohydraulic system components include the table or frame, the hydraulic power supply, a mover (actuator) controlled precisely by servovalves and the transducers to record force and motion. For axial systems, load cells and LVDTs record the linear force and motion, respectively. Additionally, for biaxial systems, torque cells and RVDTs record the rotational force (torque) and angular motion (rotation), respectively.

These systems enable precise control through feedback. Loading machines utilize a control feedback device known as a proportional-integral-derivative (PID) controller, Figure 4.4. In the simplest terms, PID controllers provide the loading machine with the information necessary to correctly "interpret" the material that is being tested. Whereas older machines often have manual settings for feedback, for example potentiometers that are set with flathead screwdrivers, newer commercial systems often have "autotune" protocols that enable automatic adjustment of the feedback settings. This is one of the major advantages of the commercial platforms and this is often where the cost is justifiable. This is particularly true if using the machine for testing elastic or viscoelastic (compliant) materials in dynamic settings.

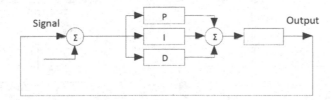

Figure 4.4: Diagram of a simple feedback loop used in mechanical testing machines.

The control mode is chosen by the user and may be load, displacement or strain. For example, if operating a machine in load control the load cell's electrical (voltage) signal is used as the feedback signal. The operator determines the desired load and rate of load application (loading rate). The controller, using the command signal from the input load signals the actuator to move. The actuator continues to move while the load cell feedback is monitored. When the signal corresponding to the user-defined input is achieved, the actuator stops moving and the load is reached. For displacement control, a similar operation occurs, except the LVDT is used as the control input and is monitored in feedback. For biaxial machines, load and displacement control extend to torque and rotation control using the torque cell or RVDT as the control input, respectively. For strain control, external sensors (e.g., strain gauges) are used to monitor feedback.

Research needs, personal preference, individual skill sets and finances are among the factors that will dictate the type and number of loading machines one will encounter in a biomechanics laboratory. Loading machines can range in complexity and price from the relatively inexpensive, in-house single screw axis driven system to the high-end biaxial and biplanar electromechanical systems. Given that the biomechanical engineer will often be trained to run a particular system or provided with manuals and video training for self-paced instruction, it is not the intention here to explain the details of operation of every make and model of machine. Instead we will focus on the basics of operation of measurement and illustrate the basics of machine development.

The purpose of a testing machine or generic loading platform as we will discuss it in this text is to provide a basic frame in which load may accurately and reproducibly be applied to an object to determine the behavior of that object under the applied load. The basic components of a mechanical/materials testing machine as we will define it are: planar (xy) motion; automated vertical (z) motion; a power source; load and displacement sensors; data collection (scanner and computer); and testing fixtures (3-pt bending). A simple platform schematic is shown in Figure 4.5.

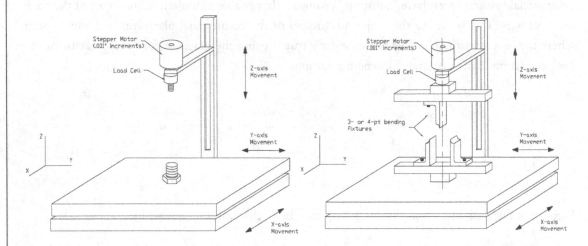

Figure 4.5: Basic components of a loading platform (left) augmented with testing fixtures for conducting three-point bending (right).

In its simplest form, a loading platform is a frame that accommodates fixtures which can be aligned to properly hold a specimen. In general, the frame provides a rigid base and a mover, to which the loading fixtures are attached. During testing one end of the specimen is attached to the fixture in the fixed frame while the other end is attached to the fixture on the mover. In this manner specimens can be subjected to a variety of testing modes including tension, compression, bending, shear and torsion.

Testing platforms should be flexible to accommodate several types of tests to maximize usefulness. Although single purpose systems (e.g., hand crank bending devices) exist, the more common approach is to purchase a flexible system or develop a basic platform. Fixtures are then purchased or designed and fabricated to work within the loading frame and enable a variety of mechanical tests to be conducted.

4.2.2 DEVELOPMENT OF A SIMPLE PLATFORM

Although there are a variety of ways to design and fabricate a loading machine here the focus is on a very flexible system that we have developed and successfully adapted over the years to accommodate a wide range of testing needs. While many laboratories develop their own systems and we are not implying that there is only one way to fabricate a system, we can discuss in detail the choices we made and the reasoning behind these choices.

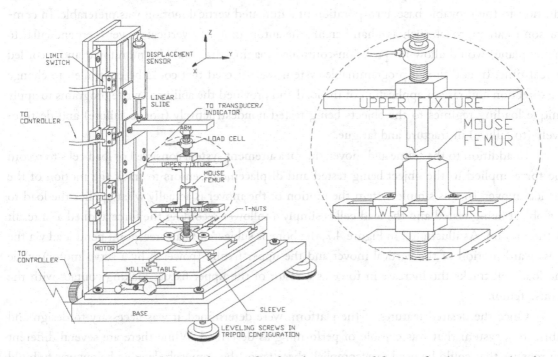

Figure 4.6: Schematic of loading platform outfitted with sensors and fixtures for three-point bend testing.

The loading machine will be referred to as a "platform" for this discussion. The "platform" as used here defines the frame, mover, sensors and data collection. The goal in this case was to build a platform to which test fixtures (tension, compression, bending) could be added to conduct biomechanical tests on the small scale (rodent limb scale).

For discussion purposes, the platform is broken down into two systems: (i) the frame and its motion and (ii) the sensors and data collection. A simple sketch of the components of the fabricated platform augmented with three-point bend fixtures is provided in Figure 4.6.

Platform Frame and Motion

It was minimally desired to incorporate in the loading platform a base frame that allowed motion in the xy plane, Figures 4.5 and 4.6. While the tests are conducted with a fixed plane, many commercial systems only allow pure axial alignment and for expanding the system (as will be demonstrated for torsion and rodent limb fracture healing studies), the ability to move the support frame of the platform in both the x and y directions increases the number of potential uses for the machine. In addition to the movable base, incorporation of automated vertical motion was preferable. In comparison to an approach such as a hand crank, the automation of a vertical mover (perpendicular to the xy plane) would allow for precision-controlled loading that could be monitored and recorded in real time. In addition, a programmable system was selected that could be controlled to change the direction and rate of application of the load and provided the ability to write programs to apply unique loading regimes to the objects being tested nondestructively (not to failure) and destructively (to failure) in fracture and fatigue.

In addition to the frame and mover, the measurement system consisted of load cells to record the force applied to the object being tested and displacement sensors to track the motion of the vertical mover. In this simple system the motion of the mover is actually what applies the load to the object being tested and the load cell is simply a follower recording the force applied as a result of the motion. As illustrated in Figure 4.7, the bone is subjected to a three-point bend load via the downward motion of the vertical mover and the upper fixture (powered by a servo motor) while the load cell tracks the increase in force as a result of the upper fixture making contact with the mouse femur.

Once the desired features of the platform were determined, it was necessary to design and fabricate a system that was capable of performing as described. While there are several different approaches that could be used to accomplish these tasks, the approach detailed below was to build the loading platform around two commercially available design elements: a milling machine table and a programmable linear slide, Figure 4.8.

Widespread use of the milling machine table (Palmgren; 100 mm by 150 mm travel) in machine shops resulted in the affordability of the table. The required planar movement accuracy (0.0127–0.0254 mm) for high precision mill work is more than adequate for this loading system. Using the milling machine table, the x and y positions can be independently dialed in to a resolution of 0.0254 mm. Once in place the positions can be locked with a set screw while etched dials enable reproducible placement. While in-service use does not require xy motion during testing, and most commercial testing platforms are used on axis, here a movable table was utilized because it simplified the alignment of the system during setup and enabled future adaptability with systems such as the torsion system discussed later in this chapter.

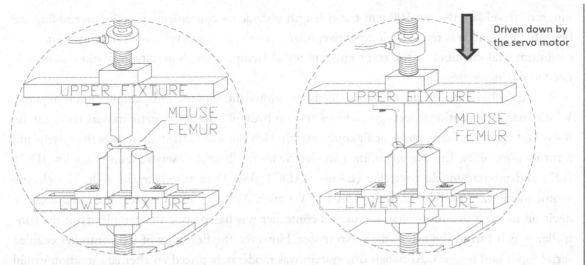

Driven down by the servo motor

UPPER FIXTURE

MOUSE FEMUR

LOWER FIXTURE

UPPER FIXTURE

MOUSE FEMUR

LOWER FIXTURE

Figure 4.7: The upper fixture moves to contact the mouse femur which rests upon the stationary lower fixture. The motor drives the downward motion of the fixture, and the load cell records contact with the specimen.

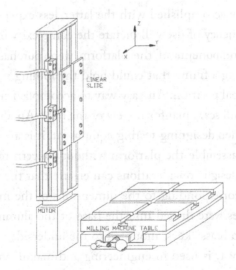

Figure 4.8: Commercial components of the loading platform.

The linear slide (Parker Daedal, 404xR100) driven by a servo motor and controller enables user-defined programmable waveforms to be input. These slides utilize a square rail ball bearing and a precision ground ballscrew drive. Depending upon the space requirements these slides can be in-line or parallel mounted. In addition, magnetic switches can be attached to an external channel on the slide for end of travel limits and home positioning. In this system a linear slide capable of 100 mm of travel and in-line mounted was utilized. Although this company provides a wide range of travel lengths the 100 mm travel length provided a convenient space for assembling the bend fixtures and was relatively inexpensive. Moreover, the larger travel enables the insertion of an environmental chamber in the event environmental (temperature, humidity, hydration) control is needed during testing.

To enable accurate, fine and reproducible motion control of the slide, automation is required. While many conventional testing machines rely on hydraulic and pneumatic movers these can be noisy and require either a pump or air compressor, which can add to the bulkiness of the system and limit its portability. This system utilized an electric motor (brushless servo motor (Exonics (IDC), B22) and programmable controller (Exonics (IDC), B8961) to automate the slide. The electric motor maximized portability to any 110/120 V outlet. Although many motors are acceptable the decision to use a brushless servo motor and controller was based upon the complexity of the controller which forced the use of the servo motor. However, the flexibility of the controller enabled serial input and triggers. Although this system was moderately priced an alternate solution would be a simple stepper motor with programmable controller such as the SmartStep 23 IDC step drive controller and the S21T step motor also available from Exonics. The use of the machine for hard tissue fracture studies generally does not require anything more complicated than a simple linear ramp waveform that is readily accomplished with the latter, less expensive option (stepper motor). Again, budget, need and frequency of use will dictate the complexity of a device.

Once the two main components of the platform were purchased, developing the loading platform reduced to fabricating a frame that could hold the milling machine table to a fixed base and the linear slide in a vertical position. An easy way to accomplish this was to lay out on a piece of cardstock/poster board a full scale planform view of the base. An aluminum base plate and vertical plate were designed. When designing testing equipment, it is always ideal to think about the ability to disassemble and reassemble the platform without concern regarding the realignment of the frame. A few additional design considerations can ensure that the platform is easily realigned. The base frame has an area corresponding to the dimensions of the milling table that is recessed. In addition, two circular holes were drilled into the base of the aluminum plate to accommodate the shape and location of two brass "keys" mounted to the underside of the milling machine table (Figure 4.9). This simple keyway is used in engineering and manufacturing frequently to connect moving components. A similar concept is used here to ensure that the base of the milling machine table is mounted to the aluminum base plate in one way and only one way. This increases reproducibility in testing by guaranteeing that the table and plate are always secured in the exact same position. Two clearance holes in the milling machine table dictated the location and size of the drilled and tapped holes machined into the aluminum base plate. With the recess, keyway and two cap screws, the milling machine table is rigidly and reproducibly affixed and properly aligned to the horizontal base plate at all times.

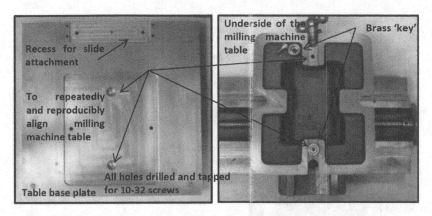

Figure 4.9: (left) Horizontal base plate with recesses for table and slide attachment. (right) Underside of table showing "keyway" system to reproducibly align table to plate.

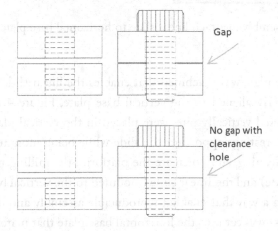

Figure 4.10: A clearance hole is used when attaching components together to eliminate any gaps that would exist if both holes were threaded.

Clearance holes are necessary when attaching materials together. For instance when you are bolting two pieces of aluminum together, if each is drilled and tapped (threaded) the two pieces will never align properly and there will always be a gap between the two pieces. To correct for this, only one of the pieces is drilled and tapped and a clearance hole is placed in the other. In this way the bolt passes through the first piece via the clearance hole and threads into the second piece bringing the two pieces together. A schematic of this is shown in Figure 4.10.

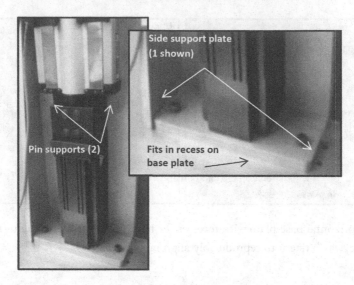

Figure 4.11: Mounting assembly to attach linear slide to horizontal base plate via an aluminum vertical base plate.

An aluminum base plate was machined that enables the linear slide to be rigidly and reproducibly affixed and properly aligned to the vertical base plate, Figure 4.11. In addition, because the slide was to be mounted vertically pins were placed in the vertical plate to relieve the weight of the slide and provide a resting support for the slide while being mounted to and removed from the plate for transport, cleaning, etc. To create the platform, the milling machine table (mounted to the horizontal base plate) and the linear slide (mounted to the vertical base plate) also had to be mounted to each other in a way that enabled reproducible assembly and disassembly and ensured proper alignment. A recess was cut into the horizontal base plate that matched the perimeter shape of the vertical plate. Clearance holes were cut through the vertical base plate and holes were drilled and tapped into the horizontal base plate to secure the two mounting plates with socket head cap screws. An additional through hole was cut into the base of the vertical plate to accommodate the electronic connection for the servo motor to the controller. Although it may not always be possible, when designing systems that need to be bolted together it is always good practice to use the same size and type of screws. The standard size is appreciated by the machinist and minimizes unnecessary risk of machining mistakes. The standard fastener size is convenient in setup in that it minimizes the necessity for multiple tools. For instance, in the system shown here a #10-32 socket head cap screw was used which enables the platform frame to be assembled with a single Allen wrench (hex key).

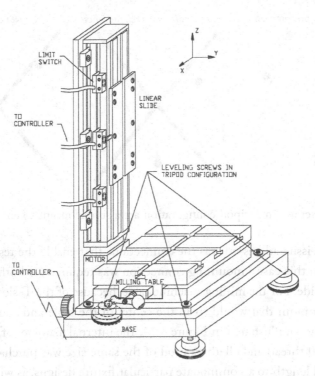

Figure 4.12: Leveling screws in a tripod configuration. Placement of screw under linear slide and motor helps to support the weight and increase system accuracy.

Another design feature incorporated into the frame was the ability to adjust the frame to maintain level. A simple solution was to add leveling screws in a tripod configuration. Using three leveling screws in a tripod configuration ensured that all three screws remain in contact with the bench on which the platform set, Figure 4.12. Designing the tripod configuration such that one support was under the vertical slide provided additional stability by acting as a jack. To further ensure that the platform sits level, digital protractors (technically, "inclinometers") are utilized whenever the assembly has been disassembled and reassembled or moved. For maximum stability, place the leveling screws equidistant on the same imaginary circle that sweeps the entire surface of the base, Figure 4.13.

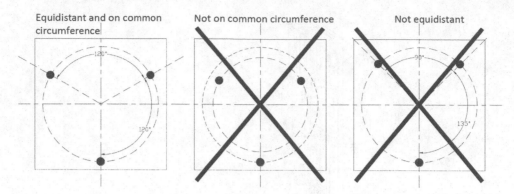

Figure 4.13: Leveling screws in a tripod configuration are a simple means to ensure a level platform.

The last design issue addressed was the connections that enable the test fixtures to attach to the platform frame. In this case a mounting connection was required for both the milling machine table and the linear slide. For the milling machine table, because of the T-slot connections a sleeve was made out of aluminum that would sit in the center of the table and could be locked in place with T-nuts. This is shown illustrated in Figure 4.14. The internal diameter of the sleeve was drilled and tapped for a ½-20 thread and all-thread rod of the same size was purchased in longer lengths and cut to the desired length to accommodate particular fixture designs, as will be demonstrated in the next chapter. In addition, an aluminum arm was designed that would mount to the linear slide for fixture attachment. The hole patterns and hole sizes machined into the arm for attachment to the linear slide were dictated by the commercial slide and were metric. Given the precision motion of the slide, we opted to work with these patterns and sizes rather than to risk damaging the slide by machining a new pattern into the moving plate. The arm, discussed in Chapter 3 was designed to hold a fixture at a fixed distance from the slide. In addition, the use of the milling machine table enabled motion to be dialed in on the table to further ensure proper alignment at all times.

The milling machine table has planar adjustability. To ensure that the upper and lower bend fixtures align in the testing machine, alignment fixtures were fabricated, Figure 4.15. As shown, the alignment system consists of threaded studs with brass collars attached to the base and linear slide. Using the xy motion of the milling machine table the two studs are brought into contact. A sleeve (toleranced to +0.025mm) is fit over one stud and when aligned will slide freely over the other stud; the table is then locked into position. This is a simple way to easily align the fixtures when testing on axis is critical. Instead of flats cut into the all-thread studs to assist with mounting, here a through hole was placed in each stud and an allen wrench can be used for leverage to break the studs loose, if necessary.

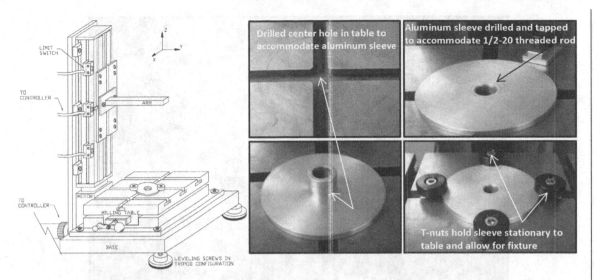

Figure 4.14: Sleeve machined to lock into the center of the milling machine table and hold the test fixturing. Given the existing t-slot assembly of the commercial milling machine table, nylon bolts with aluminum t-nuts were fabricated. The bolt is attached to the nut via a standard #10-32 socket head cap screw.

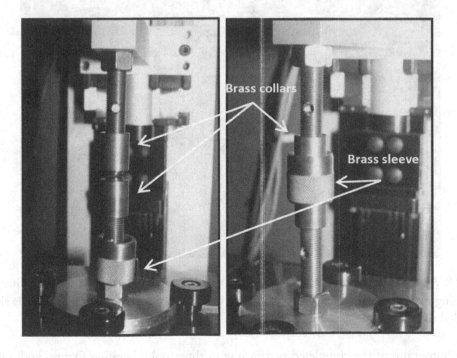

Figure 4.15: Alignment fixtures fabricated from aluminum and brass.

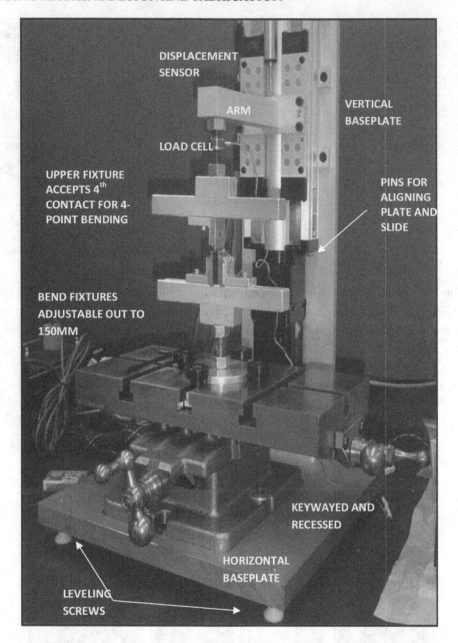

Figure 4.16: A standard platform can accommodate a wide variety of testing fixtures, conditions and uses. Loading machine augmented with three-point bending fixtures. A rodent long bone is shown being loaded to failure.

A photograph of the completed platform with three-point bending fixtures and transducers attached is provided in Figure 4.16. As will be demonstrated, a simple platform that is properly designed and machined can provide the necessary accuracy, precision and reliability that is needed to confidently conduct biomechanical testing. They key to conducting successful biomechanics testing is the effort put in prior to testing the specimen. Having a reproducible platform, developing appropriate fixtures, developing appropriate protocols, verifying the accuracy of your data via instrument calibration and the ability to utilize the instruments properly is critical. The goal is to test the specimens and let the data determine the answers. The extra effort to develop reproducible platforms and fixtures helps ensure that the data is reliable.

Sensors and Data Collection

To record force-displacement data during testing, sensors to simultaneously monitor load and displacement are required. The decision to utilize a canister load cell was based solely on familiarity with this type of system. These load cells are relatively resilient, inexpensive and their widespread use has led to a variety of commercially available ranges with high accuracy and precision. In our system we routinely utilize load cells operating in the range of 0.0625 N to 44.5 N (Sensotec, Inc, Columbus, OH). Although the displacement can be read directly from the controller into a data acquisition system, an external displacement sensor was opted for in this system. The decision was based on the ease of calibration of the sensor, electronic interfacing simplicity and the compatibility of the sensor with the existing, commercial scanner. If an external displacement sensor is desired the 25 mm sensor by Vishay (HS25, Vishay Measurements Group, Raleigh, NC) is accurate and affordable. In the current system only displacement control is available and the specimen is loaded during testing by indicating a distance that the slide will move through. Although force actuation and load control could be added, in our laboratory the majority of the intended uses require displacement control and the single control option was cost-effective.

For data acquisition, a company that manufactured the scanner, displacement sensor and strain gauges was selected. This resulted in ease of interfacing and greatly simplified the electronics of the system. Moreover, these selections simplified operation and made specialized technicians and training courses unnecessary. Vishay Measurements Group was selected because of the availability of the data acquisition scanner (Model 5100B, Vishay Measurements Group, Raleigh, NC), displacement sensors (Model HS25) and strain gauges (CEA-06-062UW-350). Although strain gauges are beneficial in biomechanics testing, if these are not needed and given that the controller provides displacement readout directly from the keypad, cheaper acquisition systems are available.

The selected scanner model has a maximum data collection rate of 10 samples per second (10 Hz). This is more than sufficient to load specimens at a rate of approximately 0.1 mm/sec to failure with a linear ramp waveform. Strainsmart software accompanies the scanner and can efficiently be navigated through online tutorials and the accompanying manual. It is also not uncommon for researchers to write their own code for data collection and analysis using Labview or similar programming software.

Calibration

Although there are ways to measure the changes in load and displacement during a mechanical test, the question becomes "how do you know that the measurement is correct?" Calibration is the process of verifying the accurate performance of your measurement equipment. In most cases this is accomplished by comparing it to a standard. For example, to manually calibrate a load cell individual weights are applied and the load cell electronic readout is directly compared to the known mass of the weights. A curve of the actual vs. expected readouts is plotted and represents a calibration curve. If the two different measurement approaches are not in close agreement, the device is not reading correctly or needs replaced. While using weights may seem relatively straightforward, these weights (Figure 4.17) are precision measurement equipment with tolerance varying by classification. Furthermore, handling and environment can affect their accuracy. Follow manufacturer suggestions in the handling and care of the weights to maximize useful life. For example, cotton gloves are worn when handling the weights to keep oils and rust from accumulating on the weights.

Figure 4.17: Precision weight sets for equipment calibration.

Linearity and accuracy of the load cell was verified using calibration weights, and the corresponding voltages (mV/V) were compared to the known weights. For the calibration example shown plotted in Figure 4.18, three weights were used: 4.45 N, 8.9 N and 22.25 N. A linear regression was fit to the experimental data and a Pearson correlation coefficient of 0.9999 was calculated. The process was repeated for unloading of the weights and an identical curve was obtained. Each time a new transducer is purchased the calibration should be verified. In addition, the process should be repeated periodically to verify that the transducer is functioning properly.

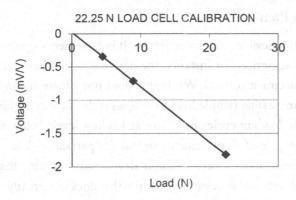

Figure 4.18: Calibration of load sensor.

Linear measuring devices, such as calipers and micrometers, can be calibrated by measuring objects of known dimensions and determining the degree of error between the values. As previously demonstrated gauge (Johanssen) blocks can be used as the calibration standards for displacement sensors. These blocks also may be used to quantify the motion of a displacement sensor and can be wrung together and/or used with feeler gauges to span any range needed.

To verify the linearity and accuracy of the displacement sensor the controller was used to run the loading arm through the sensor range with the corresponding voltages (mV/V) recorded. For comparison the manufacturer-supplied calibration data was used to calculate voltages for the corresponding displacements. The results are plotted in Figure 4.19. A linear regression was fit to the experimental data and a Pearson correlation coefficient of 0.9999 was calculated. To verify the precision, the procedure was repeated four times. Correlation coefficients for the four runs were within 0.5% of each other. Calibration was completed with the sensor initially in both the extended and retracted states.

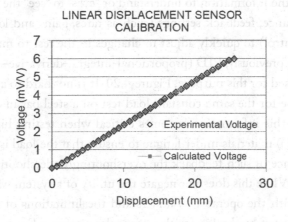

Figure 4.19: Calibration of linear displacement sensor.

Limitations of Loading Platform

While the system is highly flexible and cost-effective, it is extremely important to acknowledge the limitations of the in-house device. For instance, the device does not have feedback and as built can only be run under displacement control. While this does not negate the usefulness of this system for relatively rigid fracture testing (single load to failure) it does affect highly viscoelastic materials and the fatigue (multiple loading cycles to failure at loading levels below that inducing fracture) of these materials. In the case of displacement control the particular slide chosen is controlled by a servo motor that operates under a series of user-developed programs that control for variables such as displacement, velocity and acceleration. While this does not greatly affect a simple fracture test (using a linear ramp waveform), the device is not a convenient tool for applying a frequency driven waveform, such as a pure sinusoid. For these needs user-defined programs are curve fit to characterize oscillatory waveforms that approximate within reason (< 5%) a desired sine wave. It also requires the adaptation of protocols requiring load control to a displacement control model. However, not all protocols should be modified for displacement control. If load control tests (e.g., creep) are critical to your work, this system would need load control added.

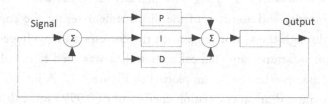

Figure 4.20: Diagram of a simple feedback loop used in mechanical testing machines.

In the case of feedback, the simplest way of envisioning this concept is that feedback provides the machine with the information to understand or "eyes" to "see" the material/specimen that is being tested. For instance, feedback settings (such as rates, gains and loops) enable a machine (running under load control) to quickly adjust to changes in the test to maintain a constant load. As briefly touched upon previously, PID (proportional-integral-derivative—and sometimes PIDO (offset) controllers are used for this purpose, Figure 4.20. It is not hard to appreciate how different this adjustment would be for the same constant load test on a steel bar in comparison to a rubber strip. And the need for this information becomes critical when testing highly elastic/viscoelastic (high degree of hysteresis) materials under fatigue to ensure that the load is efficiently reached and maintained. In the absence of such feedback the overshooting/undershooting of the load can lead to very erroneous data. While this does not negate the utility of a system without feedback, it does put the responsibility with the operator to understand the limitations of the testing system and determine if reliable data can be obtained with a particular platform. Furthermore, the system built here is appropriate for relatively rigid materials and fracture tests (single load to failure). Given that the commercial slide is not fatigue-rated, the machine is used sparingly for cyclic loading. This is because fatigue will cause wear and tear on the system that will affect its ability to function properly by introducing slop into the system.

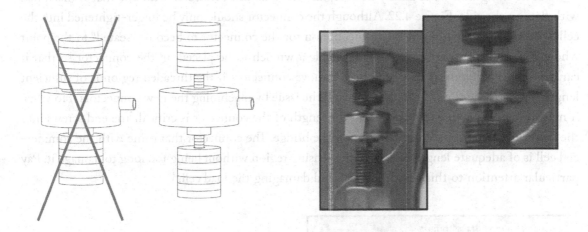

Figure 4.21: Weak link is load cell connectors that may be easily bent or sheared off on misuse or off-side loading.

Once the system has been assembled, the weak link is the load cell connectors. These threaded studs come from the company and are finger tightened into the assembly (one studded end is machined as part of the canister). As is the case with strain gauge-based transducers, canister load cells work by contact with a sensing junction on the load cell. If the assembly is overtightened the sensing junction can be overloaded. For load cells on the smaller scale of this loading machine, even slight overloads (particularly bending or side loads) can cause irreversible damage. To avoid this the load cells must be carefully tightened (finger tightened) in the studs and a slight gap should be visible, Figure 4.21. This will ensure that the cell has not been damaged during setup, but will enable slight "wobbling" of the upper fixture if bumped. To minimize the wobbling fixtures should be small and lightweight, stud connectors should be short and loads applied should be small. Thus loads on the order of 150 N should not be used on this system given this limitation; and, importantly these loads should be applied axially. If the specimen being tested is not symmetrically aligned with respect to the loading machine, the bending moment can induce large off-axis loads that can bend the load cell connector and damage the cell. Furthermore, the linear slide has a specified load that it can safely lift when oriented in a vertical position. Although this is on the order of 225 N, this load is assumed to be applied directly to the surface of the slide. Given that our system utilizes a loading fixture arm that extends out from the slide, it would not be advisable to attempt a load of this magnitude. Thus, given these two limitations this machine would not be recommended for use (as is) to impart loads over 150 N and care should be used to ensure that loading is on axis. If larger loads are desired, heavier slides can be purchased. This in turn would require load cells with the capacity to read larger loading ranges and sturdier connectors.

One thing that we routinely do is replace the load cell connector with one that is machined with flats on the side, Figure 4.22. Although the connector should only be finger-tightened into the cell, with repetitive loading it is not uncommon for the connector to become "seated" to the point where it is difficult to remove. The flats enable a wrench to be placed on the connector so that it can be removed. Jam nuts may be used to remove connectors if the threaded region is of sufficient length, but we have found it simpler to avoid the issue by machining the new connectors. However, if machining connectors remember that the length of the connector is critical. The end screws into the load cell thereby enabling activation of the bridge. The connector that came with the commercial cell is of adequate length to engage the sensing region without being too long to damage it. Pay particular attention to this dimension to avoid damaging the load cell.

Figure 4.22: Replacement load cell connectors are machined with flats to assist with removal.

As is typical of biomechanical testing systems, machine deformation is largely unaccounted for but assumed to be negligible as long as machine stiffness is much greater than specimen stiffness. For our uses, machine stiffness is greater than specimen stiffness. As noted earlier, while this system does not provide feedback, the system is acceptable for loading relatively rigid objects such as rodent long bone. To set up the machine prior to breaking rodent long bones, we commonly verify performance using wooden matchsticks. This allows us to verify that the machine is working properly, that the transducers are reading and that the loading and data collection rates are appropriate. We use the matchsticks to avoid wasting the test specimens in setup. As shown in Figure 4.23, the system generates data that is easily reproduced, and the smoothness of the data indicates that the load cell selected is in the appropriate operating range to generate reliable data. The qualitative shape is representative of a typical rigid body test and a nice linear region, yield transition and failure characteristics are clearly distinguishable. In addition, from a quantitative standpoint the results are feasible. Remember to quantitatively consider the testing data. That is, do the results make sense? If the force shown were an order of magnitude higher, they would clearly be incorrect. The question would then be is it something wrong with the testing and the testing process, or was there an error in data analysis? Data should always be looked at to verify that it makes sense from both a qualitative and quantitative standpoint.

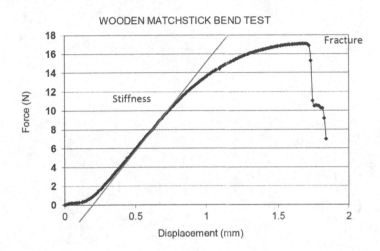

Figure 4.23: Load-displacement curve from a wooden matchstick linearly loaded to failure.

4.2.3   ADDITIONAL LINEAR APPLICATIONS

Once a platform is developed it may be expanded to accommodate a wide variety of testing needs. For example, as illustrated in Figure 4.24, the system could be expanded to four-point bending (4.24a), compression testing of trabecular and cortical bone block properties and novel tissue scaffolds (4.24b), *in vivo* modeling of rodent limbs in either anterior-posterior or medial-lateral orientations (e.g., controlled loading, exercise) (4.24c), *in vitro*, *in situ* and *ex vivo* modeling with the addition of an environmental chamber (4.24d), and tension testing (4.24e) or bend testing (4.24f) of single osteons (bone microscale structure), to name a few practical examples. Finally, the system enables both fracture and fatigue testing. However, in the event of the latter the fatigue cycle must be defined in terms of displacement and not load given that this system as fabricated only has displacement control and fatigue should be kept to a minimum because the slide is not fatigue-rated. While the system would do poorly in loading an implant to 1 million cycles, bones cycled at a physiologic load will endure far fewer cycles to failure. However, we still use it sparingly for fatigue work.

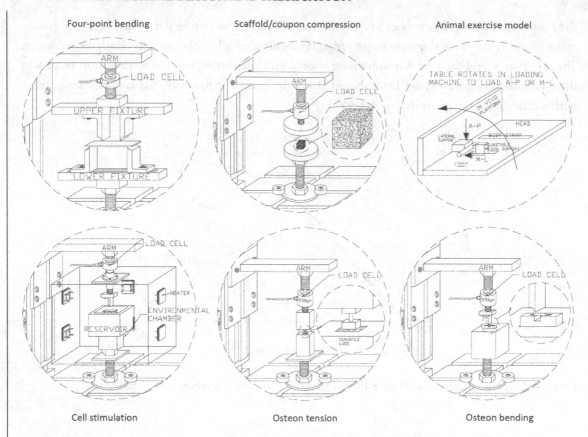

Figure 4.24: The goal of a generic loading platform is to create an accurate and reproducible loading frame that can be used to accommodate a variety of biomechanical tests. Here the system is shown outfitted with fixtures for a variety of uses.

4.3   EXPANDING THE SIMPLE PLATFORM BEYOND AXIAL MOTION

4.3.1   TORSION

As demonstrated, a simple, reliable loading platform utilizing linear motion may be readily developed. It can then be outfitted with a variety of fixtures to mechanically subject objects/specimens to tension, compression, shear, bending, etc. While the system provides a great deal of flexibility, one area in which the device is inadequate is torsion testing. Given that mechanical characterization requires multiple types of tests and that critical mechanical properties (e.g., shear modulus, G) are determined from a torsion test it is desirable to add torsion to any laboratory's testing capabilities.

Shown (Figure 4.25) is a sketch for a rack and pinion system designed for the loading platform that was previously fabricated, Figure 4.16. The rack and pinion is an easy way to produce rotation from a linear motion. It is important to preface that historically "torsion" tests conducted in biomechanics studies would more appropriately be categorized as "twist" tests. Whereas in a pure "torsion" test both ends of the specimen are rotated in equal and opposite directions, in the "twist" test one end is fixed and the other is twisted about the fixed end. Given that the convention has been to refer to these twist tests as "torsion," that convention is continued here.

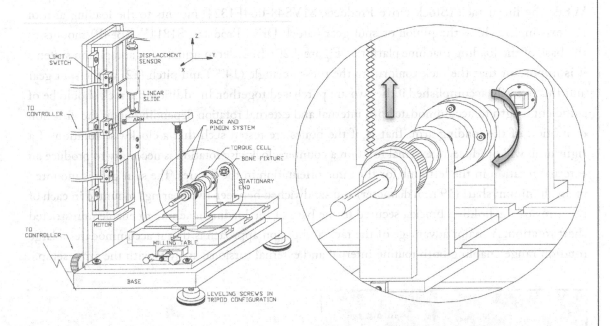

Figure 4.25: Schematic of linear loading platform outfitted with rack and pinion assembly for conducting torsion testing.

Although there are platforms developed to conduct pure torsion in biomechanics studies a limitation of these systems is that they are generally stand-alone devices that are only used for torsion testing. Here we discuss the basics of a torsional fixture that can be scaled as necessary to work in any standard single axis testing machine, such as the one previously developed in this chapter or any commercial axial system. This option was chosen because only the occasional need for torsion testing arises in our laboratory. If it is a major function of the testing lab to conduct torsion testing for material property determination it would be preferable to develop a system to enable pure torsion. When it is acceptable to compare structural properties in torsion (e.g., comparison of two fracture fixation techniques), the relative comparisons from twisting are sufficient and pure torsion is not as critical.

Initial Design Considerations

As is always the case in designing a new test fixture, before any design or machining takes place the commercial items are selected and the fixture is designed to accommodate these components. For the torsion system a torque cell, rack and pinion were purchased, Figure 4.26. Torsion may be accomplished with a variety of mechanisms incorporating sliders, cams and bell cranks. In this work a rack and pinion system which utilizes a commercially available gear set (spur gear and rack) to translate the linear (vertical) motion of the loading machine into rotational motion was developed. While the linear rack (Stock Drive Products, S1VS48-064F1324) mounts to the loading arm of the existing machine the pinion or spur gear (Stock Drive Products, S1811Y-RB-4P) mounts to the base of the loading machine platform (Figure 4.25). In order to utilize rack and pinion systems it is important that the rack conform to the pressure angle (14.5°) and pitch (12) of the spur gear and this is easily accomplished if the two are purchased together. In addition the rack should be of sufficient length to accommodate both internal and external rotation depending upon the direction of motion of the loading arm. That is, if the bones are tested such that a clockwise rotation of a right limb will produce an external rotation a counter-clockwise rotation is necessary to produce an external rotation in the left limb for the same orientation in the fixtures. The spur gear is mounted to an aluminum shaft (19 mm diameter) and sandwiched between two bearings mounted in each of the arms of the holding bracket secured to the base of the loading machine to enable unrestricted shaft rotation. A major advantage of the rack and pinion approach is that it accommodates a large rotation range enabling both routine internal and external torsional analysis with the same setup.

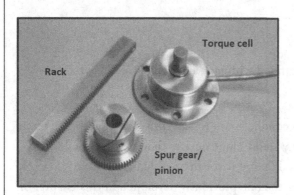

Figure 4.26: Commercial components purchased for the torsional system: rack, pinion and torque cell.

Given the existing uniaxial loading platform the goal was to develop a torsional system that would fit within this frame and would enable torques up to 176.5 Nmm (25 oz-in) to be recorded. For simplicity a torque cell that would interface with the existing data acquisition system was selected. In addition to the torque cell the 25 mm displacement sensor was outfitted to the system to record displacement that could then be converted to rotation to obtain torque-twist data in real time. The conversion is fixed for a given pinion diameter. For a linear rack movement of 25.4 mm the pinion point moves 25.4 mm along its circumference. Using the ratio of the fraction of the circumference (arc length(s)) traveled by the linear displacement to the gear radius (r), the linear motion may be determined and converted to degrees (θ) (θ=(s/r)*(180°/π)). For a pinion diameter of 35 mm, a 25.4 mm linear displacement results in a rotary motion of the pinion of 83 degrees (1.45 rads), Figure 4.27. Unlike commercial loading machines with torsion capability this system does not utilize an RVDT. Angular rotation (twist) is manually converted from the linear motion of the machine and the dimensions of the pinion/spur gear.

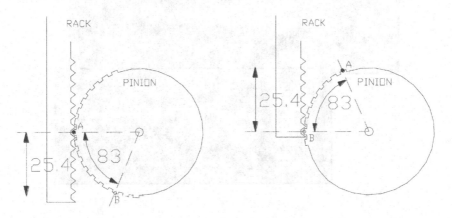

Figure 4.27: Simple mathematical relations are used to correlate the given dimensions of the spur gear assembly and linear travel to rotary motion.

Torsion System Development

Once the commercial components were purchased the goal was to determine how to attach these to the loading platform. For example, the rack has to attach to the arm of the loading machine via ½-20 threaded connectors. The torque cell which has clearance holes on the perimeter of the cell required a mounting bracket for attachment to the fixture. In addition, the torsion system once completed enabled the specimen to be twisted to failure with one end held stationary and the other moving relative to the fixed end. This required that the torque cell be mounted to the end of the fixture that directly contacts the rack. It is important to design the fixture such that the torque cell does not slide but is firmly held to the fixture and accurately records the torque applied to the specimen.

With these constraints taken into account the torsion system was designed and fabricated. Given that the scale of this system was for rodent limbs, aluminum (T6061) served as a sufficient fixture material. The rack was cut to the appropriate length for the scale of the loading machine

(100 mm) and a mounting bracket to attach the rack to the arm of the loading platform was developed from an all-thread stud that was cut to length. Two flats were milled into the stud to enable a wrench to be used in attachment. In addition, a slot and two clearance holes were cut into the stud such that the back edge of the rack (which was drilled and tapped for two screws) could be secured with two set screws to the all-thread stud, Figure 4.28.

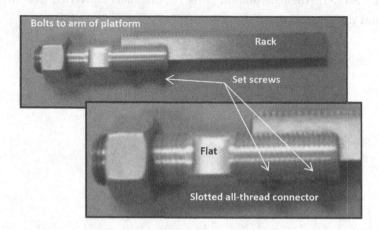

Figure 4.28: The rack was cut to a 100 mm length and attached to a ½-20 all-thread connector via a slot cut into the all-thread. Two set screws secure the rack to the connector.

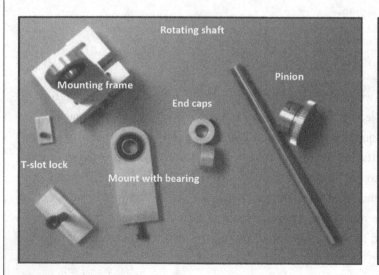

Figure 4.29: Components of the pinion system (left) unassembled and (right) assembled.

During torsion testing the specimen to be tested is mounted to the milling machine table base. One end of the potted specimen is held in a stationary fixture while the other end is attached to the fixture with the spur gear that mates with the rack for motion. The components of the fixture which interface with the rack are shown in Figure 4.29 disassembled (left) and assembled (right). The design centered around the need for unconstrained motion of the spur gear; a rotating shaft was machined from stock aluminum to hold the pinion/spur gear. The shaft is supported by a frame that consists of two vertical mounts with press-fit bearings. A base was machined from aluminum to hold the vertical mounts secure using socket head cap screws. The assembled system is shown in Figure 4.30. The frame holds the spur gear and shaft (Figure 4.30a); end caps hold the shaft in place (Figure 4.30b). Three flats machined into the shaft secure the nylon mount for the torque cell (Figure 4.30c). Set screws are tightened down against the faces of the flats (Figure 4.30d). This is important to ensure that the torque cell does not move relative to the shaft during the torsion test. The nylon material was chosen over a metal because of the need for a lightweight option, ease of machining and affordability.

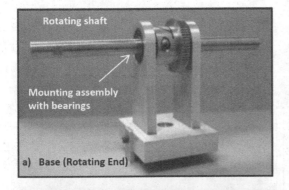
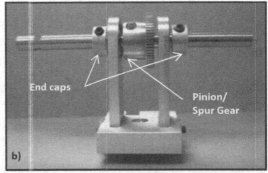
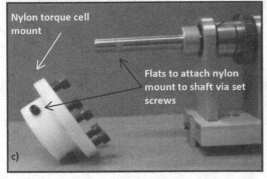
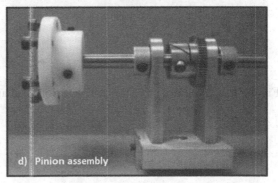

Figure 4.30: Pinion assembly mounts to the base of the milling machine table. The movable end of the specimen tested to failure is attached to this section of the pinion assembly.

To hold the fixed end of the specimen during testing a stationary fixture was also designed and fabricated. The key features of this component are the height and the ability to hold the specimen end. The height is critical to ensure that the bone/specimen is torqued along its longitudinal axis. To get a true torsional reading from the torque cell, the axis of the specimen should be the same as the rotary center of the torque cell and spur gear/pinion, Figure 4.31. The ability to keep the specimen ends from unintentionally rotating in the fixtures is also important. An easy way to accomplish this is to use square holders and potting fixtures, as shown in Figure 4.32. When designing square holes it is important to remember that these are machined with a round drill. As such, the receptacle for the specimen has rounded corners which correspond to the drill bit diameter. Although this degree of "square" is sufficient to keep the specimen from slipping during testing, if it is critical to have a completely square hole it is often easier to design this from flat plates that are screwed together. Here, the size of the specimen to be tested required use of a one-piece receptacle that was milled out to accommodate the specimen. Screws to assemble the receptacle from flat plates would be too small and would not accommodate torques as large as the single piece receptacle could accommodate.

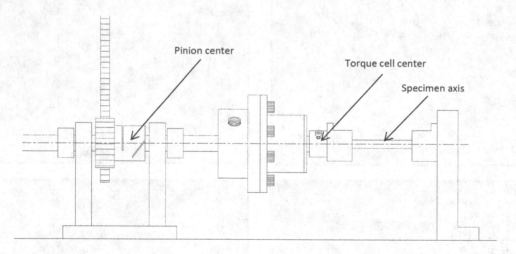

Figure 4.31: Alignment of the pinion, torque cell and specimen axis is critical.

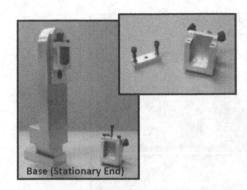

Figure 4.32: Stationary end of the specimen holder mounts to the milling machine table to hold the specimen potted in a square fixture.

The torsion system is shown assembled in Figure 4.33. The rack bolts to the arm of the loading machine while the pinion mounts to the base of the milling machine table and is engaged by the motion of the rack. The nylon torque cell mount is affixed to the shaft via the flats while the end of the torque cell has flats to accommodate the receptacle for the potted specimen/fixture. Finally, because the rack and pinion may engage in a position that does not automatically ensure the alignment of the pinion end and the stationary end of the fixture the setup block in Figure 4.33 is used to line up the fixtures prior to testing. When dealing with small objects it is particularly important to ensure that twist (pre-load damage) is not applied during test setup. This could prematurely fail the object. While careful measurement and technique could also be used, whenever possible, we prefer to develop additional tooling (such as setup blocks) that will eliminate these issues while increasing the accuracy, reproducibility and efficiency with which the test is conducted.

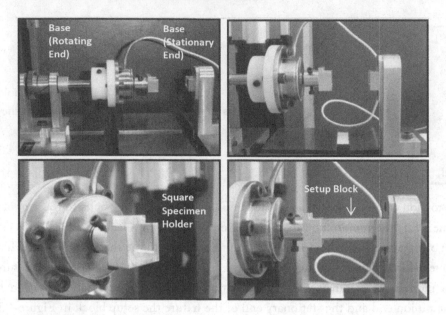

Figure 4.33: Assembled rack and pinion system mounts to existing platform and utilizes the linear motion of the slide to conduct torsional testing on small specimens.

System Performance Verification

Prior to use the system performance was verified by comparing the torsional properties of bone cement (poly(methyl methacrylate) (PMMA)) obtained from specimens tested in the system with documented properties in the literature. Using paraffin wax and stainless steel Steinmann pins a mold was made to produce uniform PMMA rods of 3.2 mm diameter. The 50 mm (length) rods were cut in half and potted in the fixtures using PMMA such that a 16 mm shaft length was left exposed. Four rods were tested to failure with torque-twist recorded, Figure 4.34. From this data, shear stress-shear strain curves were obtained and the shear moduli (G) were calculated from the slopes of the shear stress-shear strain curves. The potting technique which will be discussed in Chapter 6 adequately forced failure to the shaft and testing resulted in a spiral fracture. For the specimen shown a shear modulus of 1090 MPa was obtained. This was in agreement with previous results in the range of 1094-1100 MPa. All specimens exhibited spiral fractures. Moduli for all specimens were within 6% of each other and the testing protocol and torsional device performance were deemed acceptable. However, rigorous statistics on a large sample size were not completed due to torque cell damage concerns. That is, torque cell capacity (176.5 Nmm) was exceeded during testing. While the recorded values were actual torque cell readouts and not interpolated data, if routine testing of objects on this scale is warranted a torque cell of larger capacity would need to be purchased.

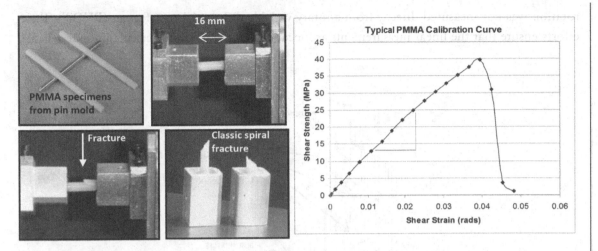

Figure 4.34: Homogenous, symmetric bone cement samples were used to verify system performance and to compare the performance of the samples in our machine to that in the literature. Performance was acceptable. While a classic spiral fracture was observed, it was not centered, largely because the test was twist and not pure torsion.

Limitations

As with any testing machine the major limitation is allowing for machine deformation independent of specimen deformation, which is seldom done. However, as with most biological materials it is assumed that if the machine is more rigid than the specimen, machine deformation is negligible and the measured deformation is the specimen deformation. An additional consideration centers around the rack. Given that the rack attaches to the arm of the platform at one end, the length of the rack is directly proportional to the compliance; the greater the rack length, the greater the compliance and the more inaccurate the results. As shown in Figure 4.35, the arm holding the rack attaches to the slide at three possible locations/elevations. Racks of different lengths were fabricated to accommodate the amount of rotation needed while minimizing compliance. For instance, for 25 mm of test travel the 50 mm rack was utilized in comparison to the 100 mm rack. The shorter rack increases system rigidity and yields a much more stable test. This is convenient when internal/external rotation is not a concern. However, if the longer rack is required the rack and pinion is engaged while affixing a dial indicator with magnetic base to the rack to verify that there is nominal deflection in the rack throughout the range of testing travel. The torsional system is also set up once and all tests are conducted with this setup to minimize the effects of setup variation. Although this may seem like a trivial issue given that twist is based upon linear rack motion it is important that rack compliance be monitored. Given the pre-existing platform, for convenience and cost concerns rotation was not measured directly from a rotary variable differential transformer (RVDT). These efforts ensure that the linear measurements are accurate and the rotational calculations are reliable.

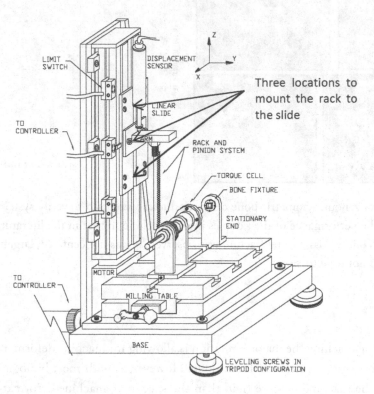

Figure 4.35: The rack can mount at three different locations on the slide with the shorter racks providing better stability. Indicators are used in setup to ensure linearity.

After addressing these platform concerns, the practical limitations of this system are then determined by the bone shaft. That is, with respect to length the shaft must be long enough that the bone ends can be sufficiently potted to concentrate the torque in the central bone shaft rather than tear the bone at the growth plate. Furthermore care must be taken to ensure that bone cement does not accumulate on the bone shaft or this will significantly alter the data. Given our interest in neonatal rat organ culture modeling we have attempted to test bones as small as 6.5 mm in length (3-day-old neonatal Wistar rat femurs) unsuccessfully. In our hands, we have successfully tested rat humeri taken from 12-day-old neonates (approximately 14 mm in overall length). The animal model is stressed in that bone dimensions may change with strain variation and transgenic alteration. Furthermore, loading rates from 0.3 mm/sec to 2.5 mm/sec were acceptable given the 10 Hz sampling rate (scanner limitation), Figure 4.36. Slower rates quickly became noisy while faster rates were insufficient to determine curve characteristics.

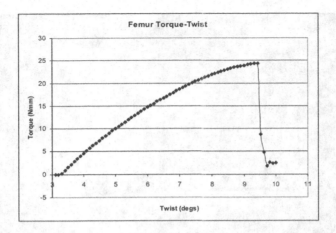

Figure 4.36: Typical torque-twist curve for the rodent limbs. Test shown here is from an adult mouse femur tested to failure at a rate of .3mm/sec.

The goal here was to utilize a pre-existing linear loading platform to accommodate torsional loading, which we felt was a practical solution to our testing needs. It is important to emphasize again that while our system would more accurately be referred to as a "twisting" system than a "torsional" system given that we are using this "twisting" system to determine structural properties our relative comparisons are still acceptable. This approach is not ideal for material property determination. Moreover, the strong agreement between the PMMA moduli and previous work should not be misinterpreted as a twist test yielding equally valid material property data as a pure torsion test. The PMMA molds were developed to test an ideally symmetric, homogenous material, a description not relevant to bone. This further demonstrates that it is important to understand the limitations of a testing system/approach. It is ultimately the responsibility of the engineer to determine which test will provide the correct information and when the equipment is insufficient to conduct the testing required.

It is helpful to think through a test prior to conducting the work. This will help you to assess the feasibility of your data. However, do not let your expectations of the results bias your interpretation of the data. If you already know the result of your study, there is no need to conduct the work. And, as you will learn with experience, it is often the case that your expectations of the results do not agree with your actual findings. This is why it is critical to develop mechanical systems to obtain accurate and reliable data. Work to ensure that your data is reliable and let your data guide your conclusions.

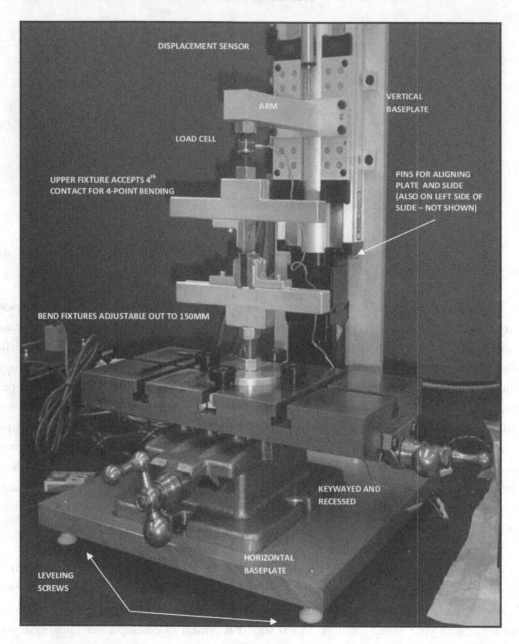

Figure 4.37: Assembled loading machine outfitted with fixtures for conducting three-point bending.

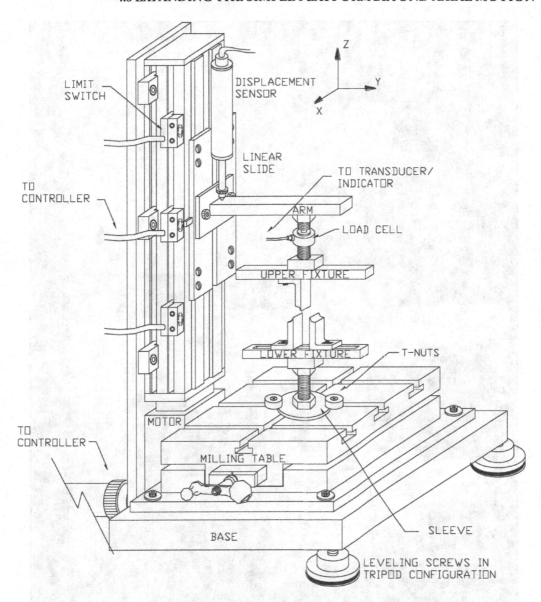

Figure 4.38: Labeled sketch of assembled loading machine outfitted with fixtures for conducting three-point bending.

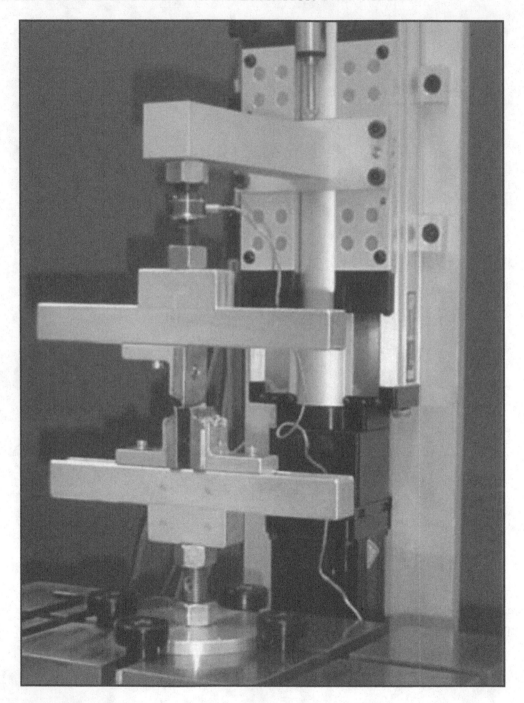

Figure 4.39: Three-point bend fixture.

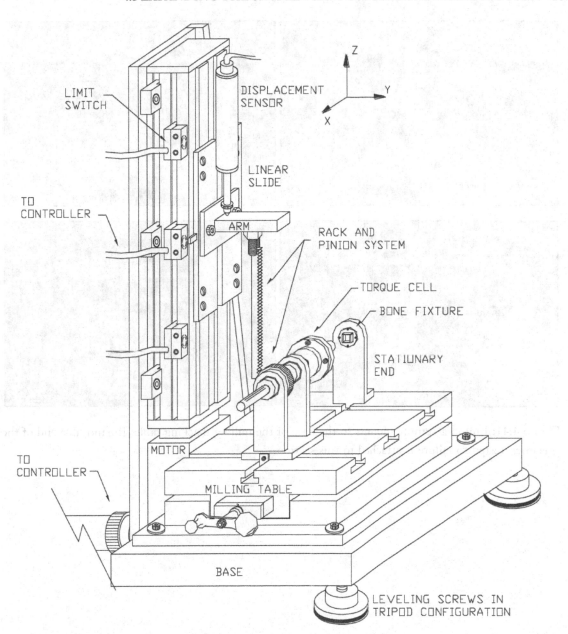

Figure 4.40: Labeled sketch of assembled loading machine outfitted with torsional platform.

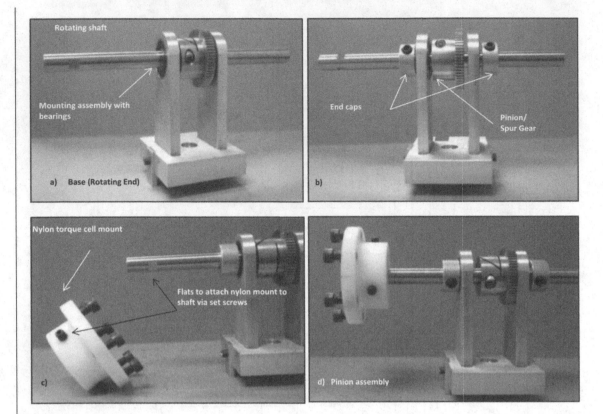

Figure 4.41: Pinion assembly mounts to the base of the milling maching table. The movable end of the specimen tested to failure is attached to this section.

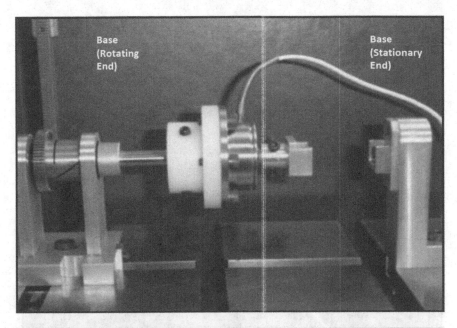

Figure 4.42: Assembled rack and pinion system mounts to existing platform and utilizes the linear motion of the slide to conduct torsional testing on small specimens which are held using square holders.

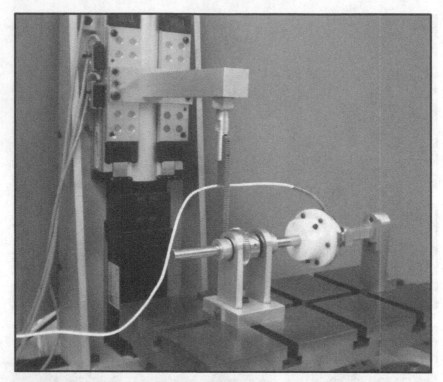

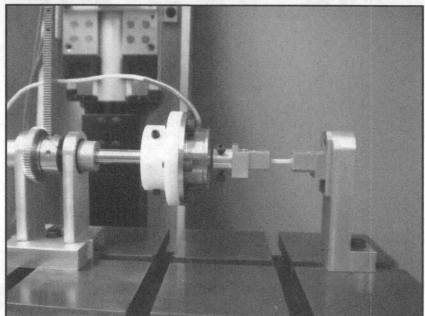

Figure 4.43: Additional photographs of assembled torsion system.

CHAPTER 5

Fixture Design and Applications

5.1 TEST FIXTURES

5.1.1 DESIGN CONSIDERATIONS

In fixture fabrication the goal is to develop a means to "hold" the specimen/object being tested during the running of the mechanical test. Although sound mechanics are at the heart of fixture design there are many ways to accomplish the same objective. This is where the creativity (and fun) comes into play.

Biomechanics laboratories will have fixtures that are specific to the testing needs of that group. While there are commercial fixtures available many laboratories have fixtures fabricated in-house or in a local machine shop. As such, fixtures will vary in appearance. Here we discuss some of the basics of fixture development and illustrate a variety of fixtures that have been fabricated and used in biomechanical testing. It should be prefaced that there are many ways that a fixture may be designed and an exhaustive account of all of these will not be provided. Fixtures for conducting tension, compression, bending, torsion and shear will be discussed.

In general, when designing a fixture you need to consider several factors including:

- What is the purpose of the fixture?

- What is the scale of the object being tested?

- What physical constraints are necessary?

- How should the object be held in/on the fixture; is there a specific orientation to maintain?

- Are there additional environmental issues that need to be addressed?

- What is the material for the fixture?

- How do you attach the fixture to the loading machine?

- Could fixtures serve multiple purposes, accommodate multiple species/length ranges?

To run through an example of this, assume that you are working with a scientist that wants to conduct bend tests on adult mouse bones. While your loading machine (and load cells) can accommodate the necessary loads your laboratory does not have the necessary bend fixtures so you have been asked to design them.

Start by thinking about the immediate needs of the fixture. For example:

• What is the purpose of the fixture?

The purpose is to conduct bend tests on adult mouse bones. If the collaborator does not insist on the type of bending you may select it. You opt to conduct three-point bend testing on the samples. Your collaborator will have a mouse model and will make the femurs and humeri available to you for testing. These bones are often selected for biomechanical testing because of the ability to create the physiologic loading of these bones using bend testing while the practicality of the relatively longer bone lengths makes them easier to handle.

• What is the scale of the object being tested?

The long bones of the adult mouse will be on the order of 25 mm in length.

• What physical constraints are necessary?

With three- and four-point bend testing, as will be shown, there is no need to physically constrain the bone ends. Since there is no physical end constraint, the physiologic relevance is not altered by the testing procedure. This is in contrast to tension and torsion testing which require physically constraining the specimen ends.

• How should the object be held in the machine; is there a specific orientation to maintain?

To simulate the bending that naturally occurs on walking the anterior surface of the bone will be placed such that the load subjects this surface to tension and the posterior surface to compression.

• Are there additional environmental issues that need to be addressed?

For a failure test of short duration it is usually sufficient to hydrate the bones prior to testing and place them on the fixture without the need for an environmental chamber.

• What is the material for the fixture?

Here the details of the system come into play. You need to consider the load cell for the appropriate loading range as well as the fixture material as the choice can add significant weight. Here a 22.3 or 44.5 N load cell is appropriate and can accommodate a heavier fixture material such as aluminum.

- How do you attach the fixture to the loading machine?

  The fixtures will need to have mounting adaptors that will enable attachment to the loading machine. On most machines this will reduce to some kind of threaded rod that can be screwed into the loading machine, both the base and the mover (ram).

Before beginning the design of your fixtures you want to address whether or not this will be a single purpose fixture or it can serve multiple functions. If it has the potential to be used in the lab for additional work it is definitely worthwhile to consider these additional uses in the design. Therefore, re-evaluate these issues with regard to additional applications of the same fixture. For example:

- Could fixtures serve multiple testing purposes?

  With few modifications a three-point bend fixture could easily be designed to accommodate four-point bending. Furthermore, there would be no additional design constraints to conduct fatigue vs. fracture testing. Keep in mind that while the machine may not be fatigue-rated, a quality fixture will hold up under fatigue loading and can be readily adapted to work in a variety of loading machines.

- Could fixtures accommodate multiple species?

  In addition to mouse limbs it is common to also conduct testing on rat limbs, rabbit limbs (we have recently been asked to test marmoset limbs), etc... While a fixture that can accommodate a human femur is unacceptable to use for rodent testing, designing the fixture to accommodate a variety of lengths enables additional animal models to be tested. For example on the scale of the mouse limb, rat long bones and larger animal tarsal bones may be tested on the same fixtures. In addition, there may be a need to conduct mechanical testing of orthopaedic hardware including pins and smaller implants. Again a metal fixture will be necessary to hold up to these loading rigors.

Although the scale of the fixture will change with the specimen model (rodent, human) keep in mind that the exact same principles apply in the fixture design. Unique challenges as a result of the scale may arise, but attention to the development of a reliable fixture incorporating sound engineering principles can overcome these challenges.

5.2 FIXTURE DESIGN AND DEVELOPMENT

5.2.1 BENDING FIXTURES

Bending fixtures are used extensively in biomechanics work to determine the mechanical properties of long bones. In addition, bending is a common loading mode to assess long bone fracture fixation treatments in cadaveric models. As such, the bones typically range from neonatal rodent long bones (femur, tibia, humerus) to human long bones. While one bending fixture will obviously not accommodate the full range of testing, an appropriately designed rodent system should minimally work equally well for neonatal mice through adult rat, in addition to small orthopaedic hardware such as Steinmann pins and K-wires. Similarly a bending fixture appropriate for human long bones could also accommodate porcine (pig), bovine (cow) and ovine (sheep), as necessary, in addition to large orthopaedic hardware such as intramedullary (IM) nails. Furthermore, it may prove useful to develop fixtures that can accommodate both three-point and four-point bending.

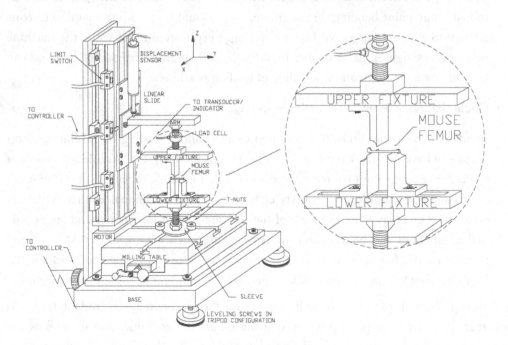

Figure 5.1: Example of three-point bend testing in a small-scale loading platform.

Once the intended purpose is determined the concepts stressed in developing the loading platform (accuracy, precision, reliability) are applied to the fixture design. That is, an adaptable fixture that can be reproducibly assembled and disassembled will ensure alignment and increase data reliability while enabling a variety of tests to be conducted. Figure 5.1 shows a sketch of a three-point bend fixture for the small-scale loading platform developed previously. As shown, the testing fixture is comprised of an upper and a lower fixture.

The upper fixture mounts to the load cell and arm of the platform while the lower fixture mounts to the milling machine table. A key feature of the fixture frame is to ensure that they are reproducibly assembled and will always align. To address alignment the upper and lower fixtures each consist of a machined channel that holds the bending contacts and an adaptor that allows the fixture to connect via a ½-20 all-thread stud to the loading arm (upper fixture) or the platform base (lower fixture). A milled recess on the top surface of the fixture for the adaptor mounting ensures reproducible alignment, Figure 5.2. In addition, two socket head cap screws fasten the adaptor to the fixture and also help to ensure that the plane is held. A "T" milled into the front face of the upper fixture and four dots milled into the front face of the lower fixture increases reproducibility. This enables easy identification of which mount corresponds to which fixture and identifies orientation front to back improving setup consistency, Figure 5.3.

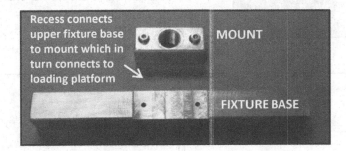

Figure 5.2: Upper fixture consists of a fixture base and the mount which attaches the fixture to the arm of the loading platform.

In addition to ensuring the alignment of the fixture relative to the platform, it is also necessary to ensure the alignment of the loading contacts relative to the fixtures. Milling a channel in the fixture (Figure 5.3) accomplishes this. Given that two screws are necessary to hold plane, with the addition of the track only one screw is necessary to hold the contact holder in place, Figure 5.4. Furthermore, the channel is drilled and tapped while the contact holder is slotted. In this way the contact holder rides in the track and the contacts can be held at any desired distance along the length of the slot. If more separation is needed the contact is moved to the next hole along the track and set to the correct distance.

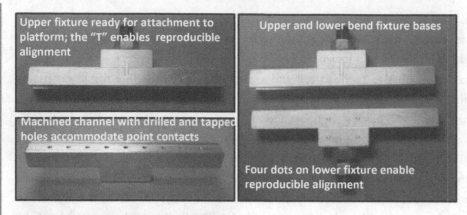

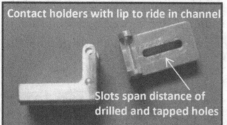

Figure 5.3: Bend fixtures consist of a contact holder that rides in the channels of the upper and lower fixtures. The series of slots in the holders and holes in the channel accommodates a span of 150 mm to enable testing a variety of specimen lengths.

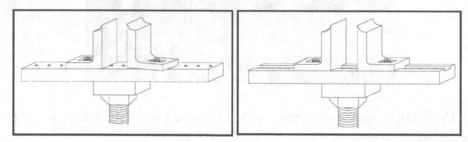

Figure 5.4: One screw is not enough to maintain alignment in the design on the left. The track design on the right enables alignment to be maintained with one screw securing each contact.

Theoretically, the bend fixtures subject the specimens being tested to a "point" load, hence the name. As such, it is important to recognize that the width of the contact where it engages the specimen requires additional attention. To accommodate a variety of different bone sizes the width of this contact was designed to be variable, Figure 5.5. Here we designed the fixtures to accommodate a round edge using a center pin held in place with spring rings (5.5 left) or a sharp blade edge with the addition of a sleeve that screws into the side of the contact holder (5.5 right). In addition, the blade edge was machined on a radius of curvature such that the blade may be ground down if damaged (nicked, bent) or a wider contact surface is necessary. Where the latter might occur is in loading (exercising) the limbs of anesthetized animals to study the response to mechanical loading.

If the blade is too sharp, it may cut the animal and damage tissue. Practical solutions to avoid this include grinding down (rounding) the blade to increase contact width, or placing rubber tubing over the blade to provide a softer contact, Figure 5.6.

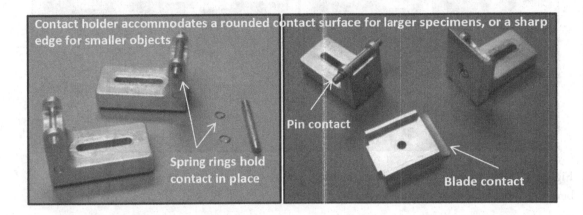

Figure 5.5: Contact holders can be machined to accommodate a variety of thicknesses to correspond to various specimen sizes and needs.

Figure 5.6: A wire's rubber sheath, cut and the wire removed can be inserted over the blade to soften the contact.

As shown in Figure 5.7, the fixtures are flexible in that they can accommodate three- and four-point bending, a variety of lengths and a variety of "point" widths. This flexibility in the design eliminates the need for several fixtures to accomplish the same tasks. In addition, the attention to detail in the machining of the fixtures results in fixtures that will align reproducibly, making them easy to assemble and disassemble as necessary (setup, cleaning) while ensuring reliable data.

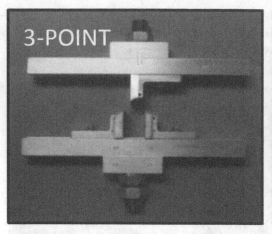
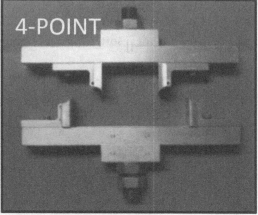

Figure 5.7: With minor modifications, the same fixtures can accommodate three- and four-point bending and a variety of specimen lengths.

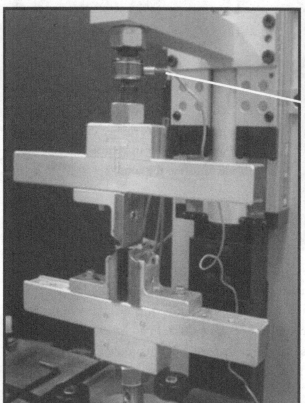

Figure 5.8: Assembled bend fixtures in platform. Weak link is exposed load cell connector that can be easily sheared without proper care and specimen placement.

The bending fixtures as assembled in the loading platform are shown in Figure 5.8, with brass blades to accommodate softer (immature) mouse long bones. Once assembled use of the system reduces to placing the specimen on the lower fixture and selecting an appropriate loading rate to ensure sufficient data collection. The key to remember in any test, bending included, is to make sure when the specimen is placed on the fixtures that it is done in such a way to ensure alignment of the specimen with regard to the loading fixture. For bending the specimen should be centered both side-to-side and front-to-back on the fixture, Figure 5.9. Do not overlook the importance of the front-to-back placement. If not on center the load recorded will be erroneous and damage to the load cell could result. For the bend test shown here, the weak link in the system is the connection to the load cell, Figure 5.8. However, this region should be exposed such that the load cell correctly reads an axial load and no side loading (moments) from unintended contact with the fixtures is present. As the rigidity of the specimens to be tested increases the importance of the alignment issue increases. If not properly aligned it is not uncommon for the load cell to shear off at the connector.

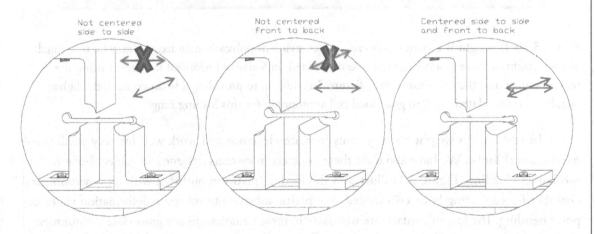

Figure 5.9: Always place specimens on center on the fixtures. Make sure they are on center with respect to both front-to-back and side-to-side alignment. In addition, if using a machine with an adjustable base, make sure the upper and lower fixtures align with respect to each other and the load cell. Fixtures, load cell and specimen must always align.

In addition to the standard fixtures shown above it may be desirable to develop alternate contacts for the smaller capacity load cells. For instance, in small-scale testing the physical weight of the fixtures may be too excessive. If this is a concern it may not be necessary to machine new lower contacts but to just machine a new upper contact. Polycarbonates/nylons are lightweight and easily machined materials that work well when using smaller capacity load cells. A variety of polycarbonate contacts are shown in Figure 5.10. They can be drilled and tapped on end to directly screw into the loading machine or they can be connected to the loading machine via an adaptor such that a quick disconnect system is created using a cotter pin, or similar assembly. The photograph in Figure 5.10 illustrates that these contacts can be used in conjunction with existing fixtures, shown here for a three-point bending test of a neonatal (5 day old) rat femur.

Figure 5.10: If weight is a critical concern, lighter, nylon or polycarbonate fixtures may be machined. Shown (right) a three-point bend test was conducted on 5-day old neonatal rat femurs using the nylon contact and the aluminum lower fixture. In addition to providing a point contact, the lighter weight accommodated the 250 gram load cell appropriate for this loading range.

In addition to weight these systems increase clearance and work well for very small (neonatal rodent) limbs. We have also used these contacts in systems designed to subject bone cells to mechanical loading. Figure 5.11 illustrates the loading platform outfitted with an environmental chamber for subjecting bone cells seeded on a plastic substrate to substrate deformation via three-point bending. The longer contacts are necessary in these situations to accommodate environmental chambers.

When testing specimens in bending it is important to address the spacing of the specimen on the contacts. Keep in mind that the amount of force it takes to break an object in bending is of little relevance unless it is accompanied by information relating to the distance between the contacts, Figure 5.12. Think of trying to bend a ruler. The closer your hands are together, the stiffer the ruler; the farther apart your hands, the more compliant the ruler. Therefore it is common practice when testing objects in bending to present the data in terms of a moment that takes into account the contact spacing. In determining contact spacing it is important to understand the model. For example, if testing fracture fixation performance in bending using synthetic bone models of identical size, a fixed contact distance for all the specimens is appropriate. The distance may be selected based upon a percentage of the diaphyseal shaft between the contacts (e.g., 80%) or it may be based upon a fixed distance from the edge of a centrally-located bone plate. If working with cadaveric models, the contact distance is modified so that the comparisons are relevant. If working with transgenic animal models, length can vary considerably across the test conditions. We often compromise when doing large populations animal studies (e.g., n=100+) and select an average contact distance for a grouping of animals (control, treated condition 1, treated condition 2...) by measuring several specimens in the group and taking an average shaft length. We then set the contacts to a fixed distance of that average length for each group. For small rodent models we often aim for a contact distance of at least 50% of the group's average shaft length. It is critical to keep track of these lengths throughout the data analysis.

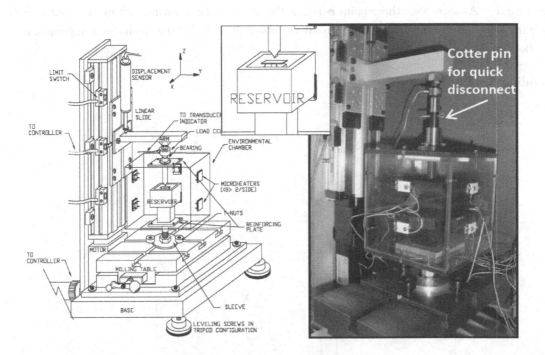

Figure 5.11: The flexible loading platform can be augmented to accommodate a wide variety of testing interests. As shown here, the loading platform accommodates an environmental chamber (controls for pH, temperature, hydration) and bend reservoir. Hydrated cells on a plastic substrate may be subjected to controlled mechanical loading and the response studied.

In the remainder of this chapter we provide several examples of fixtures and applications for their use. Keep in mind that for each fixture designed, the same attention to detail and sound use of engineering principles has been applied (alignment, accuracy, reproducibility, reliability). This section is not intended to provide a detailed description of each clinical application, but to provide enough understanding to evaluate the fixture. A few detailed projects are provided in Chapter 8.

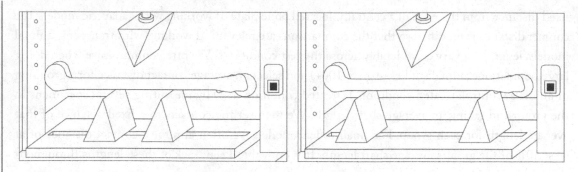

Figure 5.12: Contact distance affects the structural properties determined in a three- or four-point bending test. As shown for three-point bending, the setup on the left will result in increased stiffness over the setup on the right. Data is expressed as a moment such that the results are comparable regardless of the contact distance selected.

Bending Applications

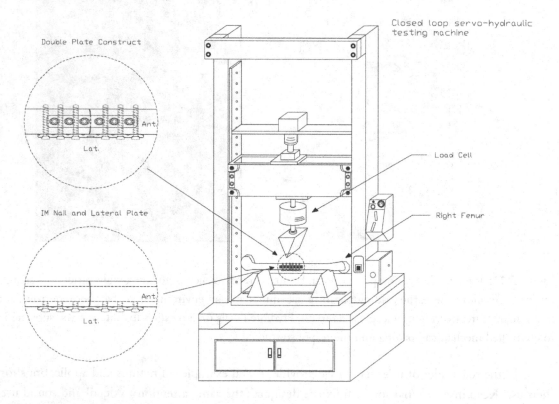

Figure 5.13: Three-point bend testing used to assess two fixation approaches for diaphyseal transverse femoral fracture fixation.

Shown here are examples of practical applications of bend testing in orthopaedic biomechanics. In Figure 5.13, three-point bending was used to assess fixation strength of transverse femoral fractures augmented with either double plate fixation or an intramedullary nail and plate fixation. The cadaveric bones were tested to failure to determine bending strength as a function of fixation technique in comparison to intact controls.

As shown in Figure 5.14, three-point bend testing with a worst-case scenario represented by the center contact in direct contact with the fracture showed that in bending the two plate construct resisted bending loads far superior to the intramedullary nail and plate combination. These results further illustrate that the two plate configuration was closer in failure load to the intact femur (control) than the combination nail and plate. This is a very routine testing setup used in orthopaedic biomechanics and is scaled appropriately to accommodate a wide range of test specimens and conditions.

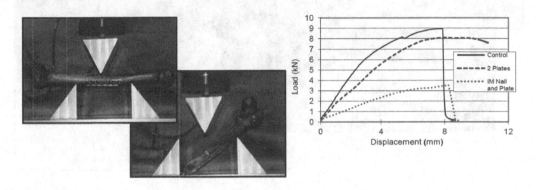

Figure 5.14: Results comparing two fixation techniques in three-point bending.

In addition to using bending in fixation studies of cadaveric and synthetic bone models, bending has also been used to create reproducible transverse fractures in animal models to study fracture healing, Figure 5.15. Basic biomechanics lends itself well to the study of *in vivo* (living animal) models. Given that in the case of fracture healing the same three-point bending setup is used, it's provided here to illustrate that point.

In a widely established fracture model, a rodent is anesthetized, the knee joint opened to expose the bones and a needle is snugly inserted into the intramedullary canal and cut flush at the bone surface. The implanted bone is then fractured; the pin remains in place to provide the stability for the animal to continue to load the bone. While methods of fracture creation vary, shown here (Figure 5.15) is a three-point bend fixture for this purpose. The system provides for a very reproducible fracture creation pattern. These studies are not limited to biomechanics but are used to assess fracture healing from a biologic standpoint. For example, shown in Figure 5.16 are femurs that have been fractured and the animals given an anti-inflammatory drug to assess the effect of the drug on bone's ability to heal. The radiographs taken at 4, 8 and 12 weeks post-fracture demonstrate the healing that took place for a control (no drug) (Ctr) and two anti-inflammatory drugs (Celebrex (Cel) and Indomethacin (Indo)). In addition to using the loading platform to create the fracture, mechanical testing is used at the endpoint of the study to quantitatively determine the bone strength post-healing. These types of studies are common in biomechanics and are generally undertaken as collaborative efforts between engineers, biologists and orthopaedic surgeons.

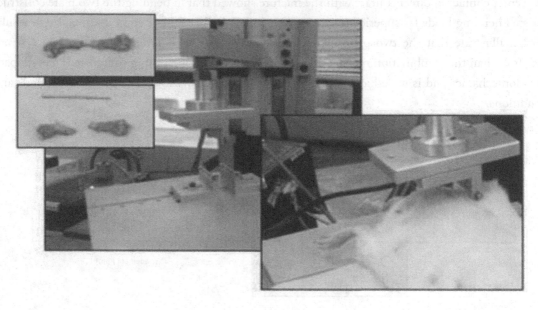

Figure 5.15: A three-point bending fixture fabricated to create stable, transverse fractures for a rodent fracture healing model.

The same principles used to design the three-point bending fixture discussed at the beginning of this chapter were used here to develop the fracture platform. Although both bend fixtures developed are for rodent limbs, here we wanted a platform base that the animal could rest upon, so we opted to develop a second system. The upper fixture is shown in Figure 5.17 with a rounded contact held in place with spring rings. The channel concept is used again to incorporate adjustment and allow for three- or four-point bending changes in contact distance as dictated by the length of the limb to be fractured and the type of fracture desired. The side to side adjustment provided by the milling machine table enables the desired placement to be precisely dialed in. This is an example where the practice of using a milling machine table that has planar adjustment simplifies setup. This orientation would be significantly more complicated to accomplish if the base of the table could not be manually adjusted. In addition to the fixtures as described, there are a variety of tools that can be developed to assist in setup. In the case of the fracture platform, the screw jacks in

Figures 5.18 and 5.19 were fabricated to prop the platform and maintain level. These simple tools were created using threaded rod and nylon caps drilled and tapped to accommodate the threaded rod. The jack is inserted under the platform and opened up to prop the platform. These are supporting structures that may be used as needed. In the fracture fixation system they help to support the weight of the fracture testing fixture and animal during fracture creation. The digital protractor used in conjunction with the jacks quantitatively verifies level (relative to the milling machine table) and is important for ensuring the necessary alignment to produce a reproducible transverse fracture.

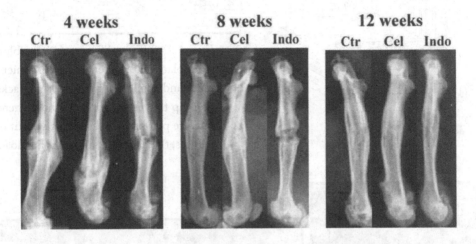

Figure 5.16: Transverse diaphyseal femoral fractures were created in a rodent model to assess the effects of non-steroidal anti-inflammatories on delayed fracture healing. Bones were mechanically tested at 4, 8 and 12 weeks post-fracture to quantify bone strength at the fracture site.

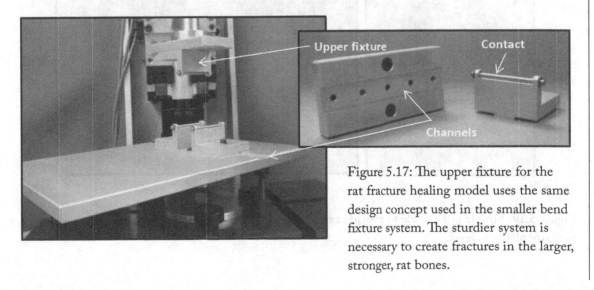

Figure 5.17: The upper fixture for the rat fracture healing model uses the same design concept used in the smaller bend fixture system. The sturdier system is necessary to create fractures in the larger, stronger, rat bones.

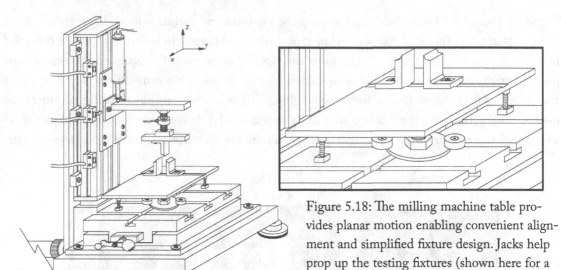

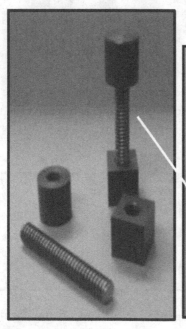

Figure 5.18: The milling machine table provides planar motion enabling convenient alignment and simplified fixture design. Jacks help prop up the testing fixtures (shown here for a fracture platform) and support weight of the fixtures, the animal and the applied loading.

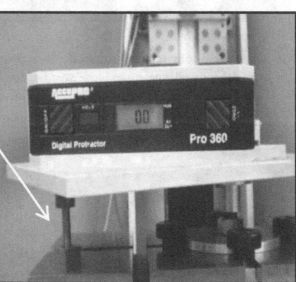

Figure 5.19: Simple jacks made to provide extra support. Digital protractors (inclinometers) are used to verify platform level.

The same concepts are used in the bending fixture in Figure 5.20 to apply exercise (isolated limb loading) to a rodent limb in the form of cyclic bending. In this example, the fixture maintains a cantilevered proximal constraint while the nylon upper fixture (Figure 5.10) applies distal bone loading. The micrometer head is used to help apply reproducible clamping given that in exercise the anesthetized animals' limbs are often loaded for brief bouts on multiple days. The fixture rotates on the platform such that the limb can be loaded in anterior-posterior or medial-lateral loading orientations. The goal in these types of studies is to subject the bone to an isolated load to study the effects of that loading event on bone cell activity and tissue level changes. Given that the limb is normally loaded in the anterior-posterior loading orientation, the isolated medial-lateral loading shown in Figure 5.20 can be controlled and its effects evaluated. These studies need tight control as they generally occur over days to weeks and the animals are then allowed to ambulate normally when not subjected to the controlled loading. The challenging goal is to tweeze out the effects of the isolated load in light of the effects of the normal ambulation and systemic effects; isolated loading in abnormal orientations can help to accomplish this.

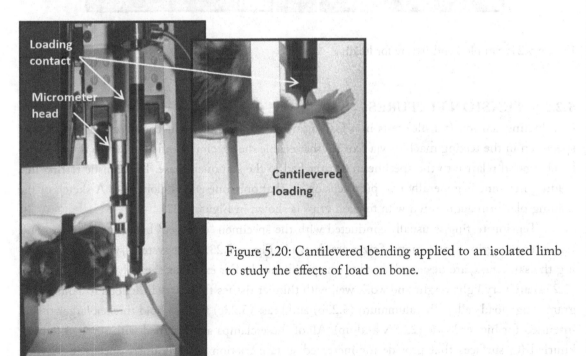

Figure 5.20: Cantilevered bending applied to an isolated limb to study the effects of load on bone.

Another bending fixture, shown here for testing a round shaft, is a simple aluminum channeled block that holds and elevates a round specimen to be broken by placing the ends in polycarbonate endcaps that rest in the channel, Figure 5.21. This system is useful for testing very rigid, small specimens and could be machined to accommodate various end shapes. This system is also very cost effective and could be mounted to a loading frame, as necessary. During testing it is necessary to ensure that the fixture does not move. Given that softer specimens could have considerable deformation during bend testing, these endcaps could overly constrain the specimens and should be avoided for softer specimens. Although simple in concept, alignment cannot be ignored to correctly utilize this simple approach.

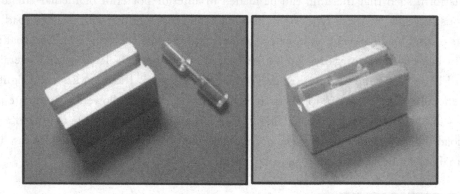

Figure 5.21: Simple bend fixture for loading.

5.2.2   TENSION FIXTURES

Conducting tension (tensile) tests in a loading machine generally reduces to securing the object/specimen in the testing machine via fixtures that enable the specimen end attached to the mover to be elongated relative to the specimen end attached to the stationary base. Thus tensile testing in a loading machine is generally not "pure tension" in that only one end is elongated. A sketch of the loading platform augmented with tension grips is shown in Figure 5.22.

Tension testing is usually conducted with the specimen "clamped" between two grips. Soft tissue tension grips for a variety of sizes are shown in Figure 5.23. These systems, while accomplishing the same task, are designed based upon specific needs. For example, the polycarbonate fixtures (5.23a) are very lightweight and work well with thinner tissues that carry little load to failure (and gram range load cells). The aluminum (5.23b) and brass (5.23c) fixtures hold thicker, larger tissues intended for higher loads (22.5N and up). All of these clamps are designed with serrated (a,b) or knurled (c) surfaces that provide for increased surface friction and minimize specimen slippage in the grip. In addition, because only one screw is used to hold the polycarbonate grips together (5.23a), the brass pin helps to hold the fixtures to the same plane and ensures that they always align, minimizing damage to the serrations. The same concept is shown in (5.23b) but scaled up to accommodate larger items that will undergo larger loads to failure. In (5.23c) a clamp has been developed that allows the specimen to be inserted without the use of wrenches and accommodates specimens of varying thicknesses. In addition, this system uses a knurled pattern instead of a serrated edge to increase the holding friction. The weight of the fixtures with the spring-loaded design works to keep pressure on the tissue to reduce slip during testing. Given the greater detail involved in the development of this fixture, additional figures are provided at the end of the chapter.

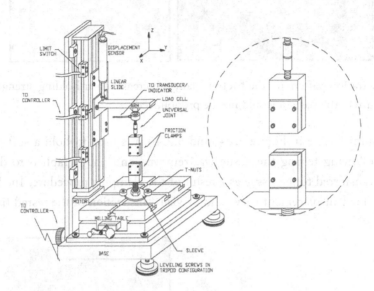

Figure 5.22: Schematic of loading platform augmented with friction clamps for gripping soft tissues (skin, tendons, ligaments, vessels) and thin specimens (mesh/stents/elastomers).

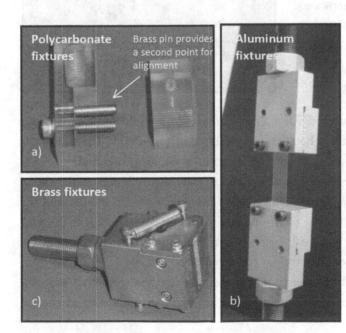

Figure 5.23: A variety of tissue grips for tensile testing. Weight is critical in fixture fabrication. Here the heavier fixtures would be inappropriate to use with gram force (e.g., 50–250 gm) load cells.

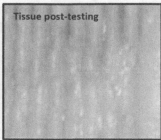

Figure 5.24: The serration pattern on the friction clamps increases their holding strength over a flat surface. If slip occurred, the pattern would not be pronounced.

The aluminum fixture in Figure 5.24 (and 5.23b) was used to hold a soft tissue specimen during testing. Following testing the tissue was removed and photographed to demonstrate the serrated pattern transferred to the tissue as a result of the clamping procedure. The lack of slip may be verified on the load-displacement curve and by visual inspection of the tested tissue.

Figure 5.25: The cryo clamp uses liquid nitrogen to freeze the tissue to the clamp. These clamps are ideal for soft tissues tested under high loads and large cycle numbers.

The tension clamp shown in Figure 5.25 is a cryo clamp used to hold muscles, tendons and ligaments for testing. The metal clamp is hooked up to a liquid nitrogen tank. The end of the tendon/ligament/muscle is tightened down with the screws and the liquid nitrogen tank hose is attached to the clamp and the valve opened. After a few minutes the tissue freezes to the clamp which prevents slip during testing. The system is convenient for longer duration studies and the tank can be turned on as needed during testing to ensure that the tissue does not slip in the grip. These types of fixtures work particularly well with larger testing machines/load cell ranges as the fixtures themselves are generally of considerable weight to prevent slip.

In addition to slip, a key issue with tension testing is ensuring that the specimens align for testing. To minimize alignment issues, we tighten the tissue in the upper grip first and let the tissue hang under its own weight before clamping the bottom fixture. Universal joints built into the upper fixture help to keep the specimen aligned during testing, Figure 5.26. In addition to helping to ensure that tension is appropriately maintained, the joint also helps to keep the load cell in line with the specimen such that the load recorded accurately reflects the uniaxial test load. Again, these concepts are equally applicable to soft tissue specimens tested in tension, regardless of size.

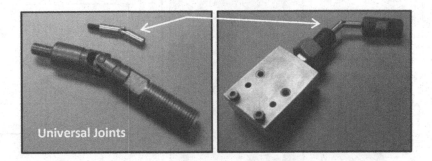

Figure 5.26: Universal joints help keep tissues properly aligned during testing. The smaller universal joint is used in the friction clamp (right).

Tension Applications

Tension/tensile testing, in addition to determining the tensile properties of a material has several useful biomechanics-related applications. For example, in Figure 5.27 a tension test is used to determine the breaking strength of skin post-healing. The simple friction clamps are more than sufficient for samples where the end effects of the clamping process are not a concern. When using friction clamps it is important that they do not cut into the material during the tightening process. If they cut into the tissue they will prematurely fail the specimen and lead to erroneous data. A common way to avoid this is to cut the samples being testing into dogbone-shaped specimens such that the failure of the object is focused to the central region of the strip where the specimen is its narrowest. This significantly reduces the likelihood of any end effects from the clamping process interfering with the testing. The American Society of Testing and Manufacturing (ASTM) publishes standards that provide standard dimensions for the dogbone-shaped specimens (ASTM D412).

There are practical exceptions to this rule. For example in Figure 5.27, while the tissues are not cut into dogbone-shaped specimens, end effects concerns are minimized given that the tissues tested here are all post-wound healing. That is, they were cut and allowed to heal before being harvested and tested. Given that the post-wound region is centered within the grips, the weak link is the centrally located wounded tissue. Relative differences in strength among the wounded tissues with different types of post-wound treatment may be successfully determined with the simple clamps and the tissue cut into constant width strips. In addition, the load-displacement curve accurately represents a typical soft tissue response, Figure 5.27. Figure 5.28 further illustrates the relevance of the dogbone-shaped specimens for intact, homogenous specimens.

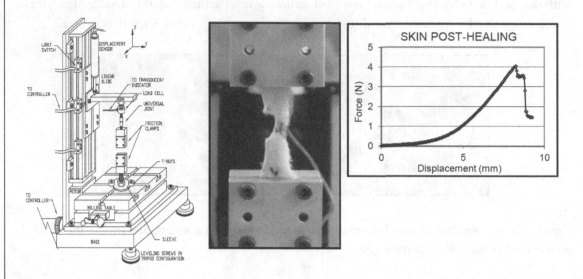

Figure 5.27: Tension testing of skin section post-wound healing. Typical load-displacement curve.

A similar example is shown in Figure 5.29. Here, tensile testing is used to assess the mechanical strength of a porcine adhesion created between the pelvic side wall and the uterine horn. The surgically created adhesion was developed with the purpose of creating a standard model by which adhesion treatment efficacies could be evaluated. Here mechanical testing is utilized to characterize the performance of the adhesion and determine baseline mechanical data that can be used to quantitatively assess novel treatments by comparison. As in the previous example, given that the weak link is the adhesion (allowed to develop over three weeks), the need to cut the samples into dogbone-shaped specimens is not necessary as long as the adhesion is centered in the fixtures for testing. While cutting samples into the appropriate ASTM standard sizes and shapes is necessary for material property determination of samples, this works best when the material is initially of uniform thickness and constant cross section. The adhesion model, shown in Figure 5.29 varies considerably in thickness and cross section of the sample. Here the goal is to determine the structural properties of the adhesion complex for relative comparisons.

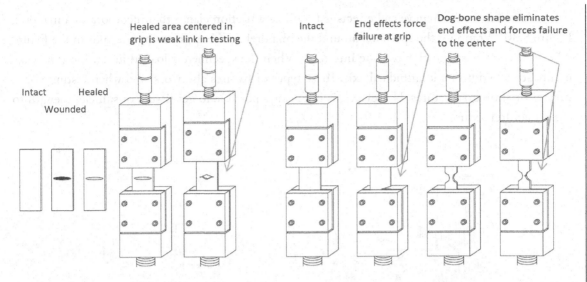

Figure 5.28: When testing specimens in tension it is important that the fixture does not prematurely fail the specimen. When specimens are modified (e.g., cut and allowed to heal), the weak link is the tissue and not the fixture interface and a constant thickness can be used. When studying homogenous, intact tissues care must be taken to insure that the clamping process does not prematurely fail the specimen. Dogbone-shaped specimens can relieve the end effects by concentrating failure to the center of the specimen.

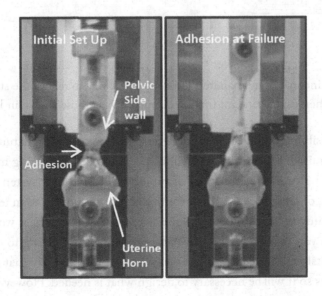

Figure 5.29: Tensile testing of a surgically created porcine adhesion.

Another way to relieve end effects is to utilize a friction clamp that incorporates a mandrel, Figure 5.30. Wrapping the specimen around the mandrel and then clamping the ends in the fixture relieves the stress at the clamp/tissue interface. When the specimen is loaded in tension it is maximally stretched on its longitudinal axis. These types of fixtures also work well when testing fibers, sutures or similar structures. However, it is not always possible to get tissues of sufficient length to use a mandrel fixture.

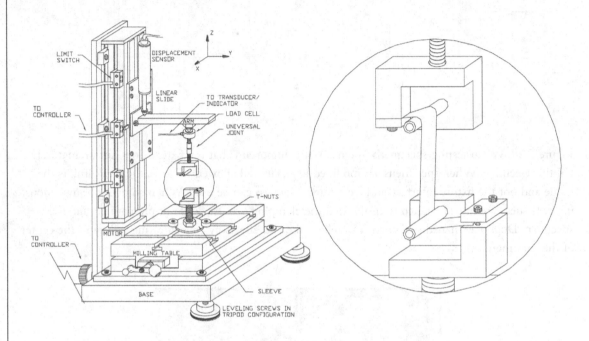

Figure 5.30: A fixture incorporating mandrels is a convenient way to relieve the stress concentrations at the grip interface when using standard friction clamps such as those shown in Figure 5.26.

It is the responsibility of the engineer to determine how to design fixtures and test protocols to obtain the data required. Oftentimes this will not be limited to obtaining material properties in standard tests but will also include a variety of projects assessing new system design and performance, quality control of new procedures and comparison of surgical fixation techniques. Also keep in mind that if the study being conducted has been previously done, there would be no reason to repeat it. The point is you may find yourself conducting tests and needing to develop fixtures that have not been previously fabricated for your exact purpose. It is unlikely that commercial systems will exist in these cases so it will be necessary to design what is needed. However, it is often the case that existing fixtures can be used in combination and it is important to remember that it is seldom necessary to recreate the wheel. Here we show a few examples in which fixtures are used in various combinations. For example, Figure 5.31 depicts a pull-out test to determine pull-out strength of a spinal screw in a synthetic bone block model. Although this is similar to a tension test, the friction clamps discussed previously are clearly inadequate to hold a screw or the block into which it is inserted. However, the same concept may be used. That is, we developed two fixtures, one which served to hold the block and one that served to hold the screw. For flexibility, given that the block size may change, a simple way to clamp an object that has parallel faces is to use a vise. A vise was fabricated consisting of an aluminum base, steel screws and knurled brass contacts, Figure 5.32. The vise mounts to the loading machine through the slotted base of the vise into which a bolt is passed and screwed into the load cell. In addition, the vise was machined to accommodate two cross rods that stabilize the vise (if necessary) during testing. For the attachment of the screw to the loading platform, a commercial universal joint is mounted to an aluminum connector drilled and tapped to accommodate the thread size of the screw (this is a great place to use a nut and bolt gauge to determine sizes). If knurled ends are not used, sandpaper can be added to increase contact friction.

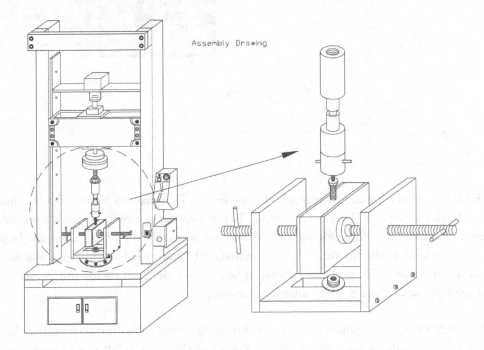

Figure 5.31: Biaxial, servohydraulic loading machine outfitted with a vise fixture to hold synthetic bone blocks for pullout testing of spine screws.

An exploded view of the system is shown in Figure 5.32 with a similar vise developed for the small-scale loading machine (photograph). For testing, the screw is placed in the block but it is easy to damage the interface (what is being tested) if the construct is screwed into the loading platform. This issue can be avoided by designing a quick disconnect system into the fixturing. Pins (cotter pins) are a good way to connect these systems. When working with pins, make sure that the fit is snug. Avoid adding any slop to the system which results in relative motion between the connections. This will introduce error into the test. Any test can be significantly hampered by slop between connections; an obvious, extreme example of this is a fatigue test where the error is compounded over time.

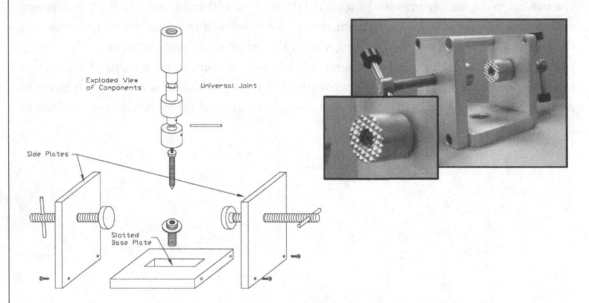

Figure 5.32: Exploded view and photograph of a vise fixture. These fixtures are very flexible and can be used to hold a variety of objects during testing. To bite into the specimens the ends are knurled (textured) to increase the holding friction. In addition, bars across the top of the fixture can be used to strengthen it for large loads. The sketched vise was designed for any fixed axis machine so a slot was incorporated for minor adjustments. The photographed vise was designed for the small-scale loading machine with adjustable base so a clearance hole was used.

Figure 5.33 shows a pullout test for a prototype ligament anchoring system. The vise fixture is used to hold a synthetic bone block while a three jaw chuck on an all-thread stud mounts the screw to the loading machine. The three jaw chuck is the same mounting system used on drill presses, lathes and mills and is a cheap and easy way to accommodate a variety of screws/fasteners. A similar system is used in Figure 5.34, but on a smaller scale. Here a commercial dental screw used for anchorage is tested. Clinically the purpose of the screw is to serve as a post that can be used to lasso and move teeth. In this case, failure of the system is not the physical fracture or pullout of the screw. The success of an anchorage system depends upon its ability to resist even the slightest motion, the failure curve here represents the initial elongation (linear stretch) of the fishing line followed by onset of screw micromotion. While the curves (Figure 5.31 and 5.32) are orders of magnitude different the same testing principles apply.

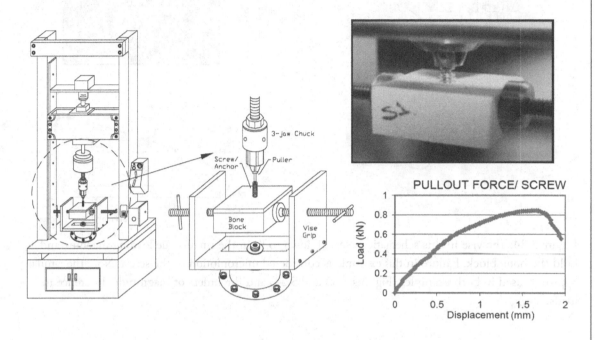

Figure 5.33: A vise for screw pullout testing is placed in a hydraulic testing machine with large load capability. A simple three-jaw chuck is modified with an all-thread stud to fit into the ram and makes a convenient fixture to grab the screw during pullout testing.

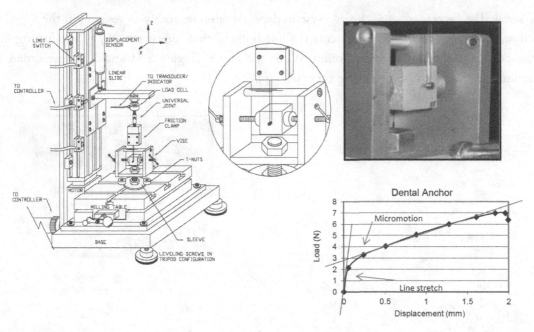

Figure 5.34: The vise in this schematic is scaled down to fit in the small-scale loading machine and hold the bone block. Failure in this example is considered micromotion of the screw. Note the similar approach used in both examples (Figures 5.33 and 5.34) and the orders of magnitude difference in loads.

5.2.3 COMPRESSION FIXTURES

Conducting compression (compressive) tests in a loading machine generally reduces to securing the object/specimen in the testing machine via fixtures that enable the specimen end attached to the mover to be shortened (compressed) relative to the specimen end attached to the stationary base. Thus compressive testing in a loading machine is generally not "pure compression" in that only one end is actively loaded.

Platens are flat plates designed to load the specimen uniformly across its entire face, Figure 5.35. Alignment is a critical issue with compression. This is particularly important for relatively brittle materials, such as bone. If the platen engages the specimen on an edge and not uniformly across the surface nor on-axis longitudinally, a moment may be induced that can cause error in the load cell readout, as well as cause concentrated load that can prematurely fail the specimen. Spring-loaded systems have been proposed that allow the specimen to seat against the platen across the specimen surface.

Shown in Figure 5.36 is a commercial swivel that was purchased as the basis of an adjustable compression platen system. The swivel shaft is threaded and a drilled and tapped all-thread stud is an easy way to connect the swivel to the loading machine. The swivel end holds a fabricated aluminum adaptor via a dovetail connection. A spring-loaded screw moves with the swivel and clamps tight against the adaptor to hold it in place, post-alignment. As has previously been demonstrated, many of our fixtures and systems are designed around commercial components. When our design involves commercial components, these pieces are always purchased first.

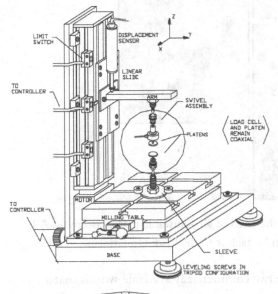

Figure 5.35: Schematic of the loading platform outfitted with compression platens for testing small scaffolds, bone coupons, and rodent vertebral bodies, as shown.

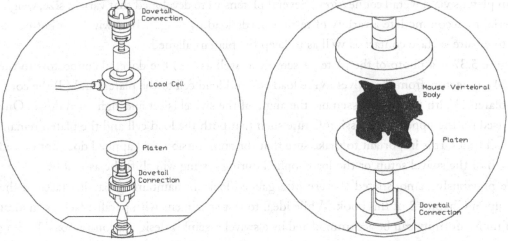

Figure 5.36: Swivels can be purchased that serve as the basis of an adjustable platen assembly. All-thread adaptors were machined to hold the swivel and attach it to the loading platform.

The assembly shown in Figure 5.37 uses the two commercially available swivels that attach to aluminum platens via dovetail connections. Several platens were developed that vary in size, weight and material to accommodate a variety of testing needs/loading ranges. The swivels on both ends are used to ensure surface contact as well as to keep the platens aligned.

Figure 5.37 is a photo of the platen assembly as well as (a,c) the dovetail connectors to the swivels, (b) connectors from the swivel to the load cell and load cell to the platen, and (d) the compression platens. With the swivel assembly the angle of the swivel is set and tightened down. One swivel is used on the upper compression fixture such that both the load cell and the platen remain coaxial at all times. It is important to make sure that the compression load applied does not exceed the strength of the swivel setup, or the load applied during testing will shift the assembly.

We previously demonstrated the use of a gauge block in maintaining parallel faces in the platen setup in Chapter 2 of this book. While ideal to test specimens with parallel faces, a modest degree of nonuniformity can be accommodated by a swivel assembly design as long as axial loading is effectively maintained.

Figure 5.37: Photographs of the compression platens machined from aluminum and attached to the loading machine via dovetail connections to commercially available swivels. An aluminum connector attaches the load cell to the upper swivel assembly. The swivels are loosened and a gauge block is placed between the two platens. The swivels are tightened down with the gauge block in place to set parallel faces. The specimen to be tested is then inserted. Again, it is important to make sure that the platen does not fit flush to the load cell or erroneous readings will be recorded. Allow for a gap on each side of the load cell connections.

Compression Applications (includes bending induced by compression)

Compression/compressive testing has several useful biomechanics-related applications. In Figure 5.37 compression platens are used to determine the compressive strength of a demineralized cancellous bone block. As with tension, compression testing is used in a variety of orthopaedic biomechanics tests. Here we highlight a few of these and demonstrate how combinations of fixtures are also used in these types of studies. In addition to compressing coupons of bone/scaffold materials, compression on the end of a bone oriented along the long axis of the bone can introduce bending, Figure 5.38. Often in biomechanics we have modeled systems in which the goal is to study this type of loading scenario. Given that the load applied to the bone end is compression, we address these examples in this section. Note however that the bone bends in the same manner as when it is subjected to three-point bending because of geometry and curvature. An often important difference is that in the case of three-point bending the bone shaft is physically loaded to induce bending. With end compression the ends are compressed and given the natural curvature of the bone, bending is induced without direct contact with the bone shaft.

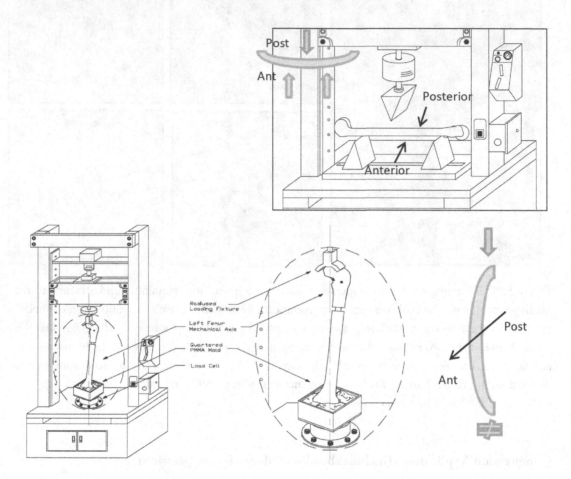

Figure 5.38: Bone may be subjected to bending in a variety of ways, including three-point bending and a compressive load on the proximal end of a longitudinally aligned bone. (Below) Bending is induced by applying a compressive load to the head of the femur. This loading is more natural in that *in vivo* the femur bends from loading that does not require direct shaft contact. (Right) Bending is induced by three-point loading. Here the central shaft (diaphysis) of the femur is under direct contact loading. This type of loading is very easy to reproduce in a lab and does not require end constraints. Note the gray sketches show that both scenarios induce the same bending.

Outerbridge-Kashiwagi (OK) Method is a procedure in which a hole is drilled through the humerus to relieve pain associated with elbow osteoarthritis. The procedure centers around the removal of a bone plug (using a trephine) in an attempt to remove pain that occurs on impingement during elbow bending. The consequence of bone loss strength as a result of bone removal is of concern. Here synthetic sawbones (and cadaver, not shown) were tested as either intact, with a 16 mm or 20 mm trephine hole. The question in this case was not whether the hole induced a stress riser, it did, but whether or not there was a functional effect. Here bones were strain gauged in the metaphyseal region and loaded to failure in compression/bending simulating a fall. In this case, the goal was to look at a worst-case scenario and assume that only the tip of the bone (point load) was contacted during loading, Figure 5.39.

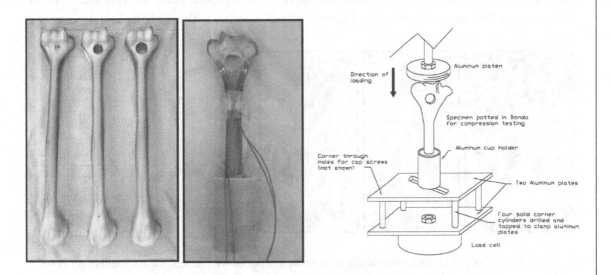

Figure 5.39: Compression testing was used to determine if a hole placed in the distal humerus increased the likelihood of fracture. Two hole sizes were compared to intact synthetic models. In addition to the loading, strain gauges were placed on the metaphyseal region to record strain.

Mechanical testing consisted of a compression load applied to the distal end of the humerus. To minimize sliding of the bone, the bone end was potted in bone cement. The loading fixture consisted of two plates with the bottom plate mounting to the load cell and the top plate mounting to the test specimen. The plates were attached via four connecting rods at the four corners and the space created by the rods provided the physical space to tighten the potted specimen in place for loading.

As shown in Figure 5.40, the models with the OK method created with the 16 mm trephine did not unintentionally create a structural weakness. In contrast the 20 mm trephine created a hole that sacrificed the structural integrity of the bone as failure occurred through the hole. Similar results were found in the cadaveric models.

As noted, physiologic bending can be induced by orienting the bone longitudinally on axis and restraining the distal end while applying a compression load to the proximal end (head) of the femur. Orthopaedic surgeons we have worked with have preferred this orientation to a three-point bend test as inducing bending without diaphyseal contact is more physiologically relevant (natural). An example of where this type of setup has been used is shown in Figure 5.41 for a subtrochanteric fracture study. While placement of the intramedullary (IM) nail in this type of fixation can be made either antegrade (from the top) or retrograde (from the bottom), there is question as to whether the construct stiffness is a function of the approach. To determine this, a variety of nails were placed either antegrade or retrograde with a subtrochanteric fracture created at one of three locations. The treatments conditions are illustrated in Figure 5.42 and correlate with standard clinical treatments.

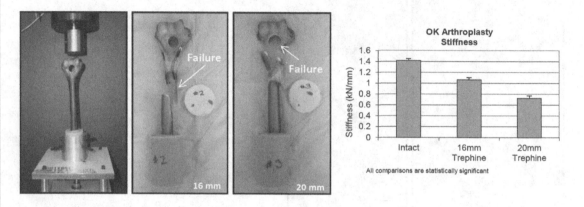

Figure 5.40: Holes drilled in mechanically composite sawbones were tested for their effect on bone fracture in compression. The largest hole size (20mm) structurally compromised the bones as fracture occurred through the hole.

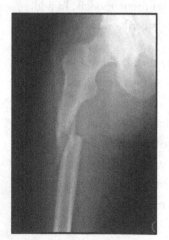

Here sawbones were used and a transverse subtrochanteric fracture (1cm width) was created. Statically locked IM nails were placed in the shafts and fixed with screws. Because the placement of the screws was blinded, fluoroscopy was used to locate the screw holes and insert the nails, Figure 5.43. The way to insert a screw into a blind hole is to locate the screw hole under fluoroscopy such that it appears perfectly round and drill straight into the hole. This ensures a straight alignment.

Figure 5.41: X-ray of a subtrochanteric femur fracture.

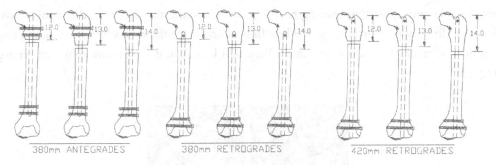

Figure 5.42: Subtrochanteric femur fractures at three locations were augmented with either 420mm or 380mm length intramedullary nails placed either antegrade or retrograde.

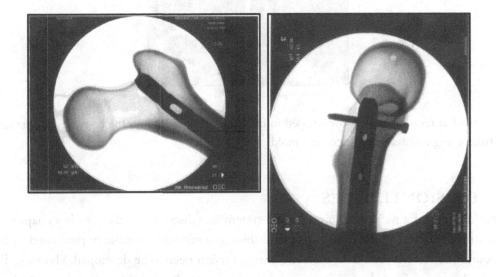

Figure 5.43: This study investigated the effects of antegrade (from the top) vs. retrograde (from the bottom) intramedullary nail placement in the treatment of subtrochanteric fractures. Given the fixation comparison nature of this study, synthetic bones served as the model. Following creation of the fracture, bones were fixed with IM nails. Fixing fractures often requires blind placement. In these cases, the rods are inserted and the screws placed under fluoroscopy (x-ray). To place the screw, the bone is rotated such that the holes appear round. Any malrotation will affect the roundness of the hole and the screw will not line up.

Once implanted, the bones were tested in the loading machine, Figure 5.44. The bones were tested by applying a preload of 40 N to the femoral head to maintain testing alignment during initial loading. Nondestructive tests were performed under load control at a rate of 150 N/sec and loaded to 1000 N. The load was selected based upon literature. Given the applied compressive load induced a bending moment and given the consistent dimensions of the synthetic composites and the reproducible and consistent loading resulting in a constant moment arm, the compressive load recorded was directly proportional to the bending moment. Stiffnesses were statistically compared as a function of fracture location, intramedullary nail length and placement approach (retrograde vs. antegrade).

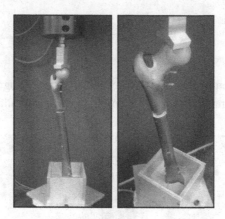

Figure 5.44: The femoral head is compressed to induce a bending load on the bone shaft. The distal end is fixed using a reusable bone cement mold.

5.2.4 TORSION FIXTURES

The development of a rack and pinion torsion system was discussed in the previous chapter for an axial loading machine. Commercial systems in biomechanics are commonly purchased as biaxial systems and the fixtures that hold the bone during torsion need to be developed. Shown in Figure 5.45 is a schematic of a cadaveric femur in a large biaxial loading machine. The photo on the right shows the fixture receptacles designed to hold the bones. They are created from four side plates and a base that connects to the plates and either the base of the loading machine or the crosshead. The upper and lower fixtures are identical.

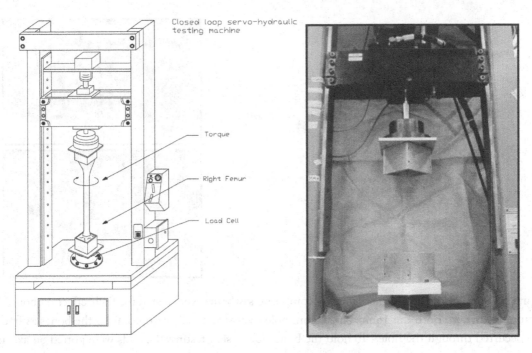

Figure 5.45: Torsion testing can be accomplished with modifications to linear loading (single axis) machines or using commercial biaxial testing machines. The load cell for this system is a combination load/torque cell.

Torsion Applications

Applications for torsion testing are numerous in orthopaedic biomechanics. Oftentimes the torsion test is not run independently of other tests. For example if an implant fixation technique in the femur would see significant bending and torsion on loading, it would be appropriate to minimally load the instrumented femurs in both modes consecutively, if not simultaneously. Examples of when torsion tests have practical applications are shown below. Consecutive tests enable the effects of each loading mode to be studied individually, as well as significantly simplify the testing process in comparison to combination loads.

As was previously noted, bones are subjected to complex loading states that combine tension, compression, shear, torsion, etc. To study the effects of load, the modes are often simplified and studied in isolation. For example, in addition to the surgeon's concern regarding the effects of the hole in compression loading on bone fracture following the OK method, torsional loading was also of concern. To address torsion, synthetic humeri were potted in bone cement (PMMA) and subjected to torsion loading, Figure 5.46. As shown, all torsional fractures involved the distal end of the humerus, regardless of hole presence or size.

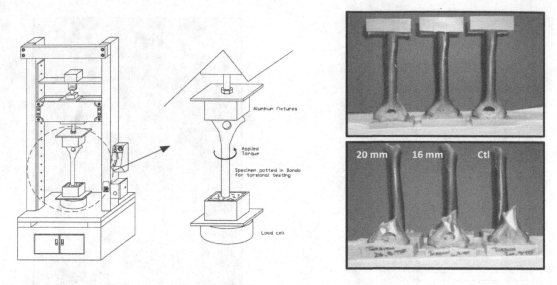

Figure 5.46: Holes drilled in mechanically composite sawbones were tested for their effect on bone fracture in torsion. Both the 16mm and 20mm hole sizes structurally compromised the bones as fracture occurred through the holes. To hold the bones for torsion testing, the ends were potted on axis in bone cement.

In a procedure similar to the OK-arthroplasty procedure, holes are placed in the distal femur to access cancellous bone that may be used to pack a fracture site when bone is needed. Again, the question is not whether the bone is compromised by the creation of the hole and the removal of the bone; theoretically it is. The question is whether or not the hole and subsequent bone removal functionally affect bone strength and increase fracture risk. In other words, the bone is mechanically compromised, but is it compromised to the extent that the procedure causes failure and therefore should be avoided? Because the fundamental equations for torsion indicate that the greatest torque is on the surface, a torsion test is an excellent way in which to determine if alterations including surface alterations affect strength. Here the same size hole and cancellous bone volume were taken from one of three locations on the femur, Figure 5.47. The bones were potted in bone cement and torqued to failure. Not unexpectedly, the more distal the hole, the less likely it is to affect structural stability of the femur, Figure 5.48. These types of studies provide feedback for the surgeons to help them determine whether a particular procedure is safe to perform and to provide data on how to maximize safety if the procedure is performed. The use of cadavers incorporate the worst-case scenarios immediately post-harvest given that *in vivo* bone can refill the hole and reduce the stress riser.

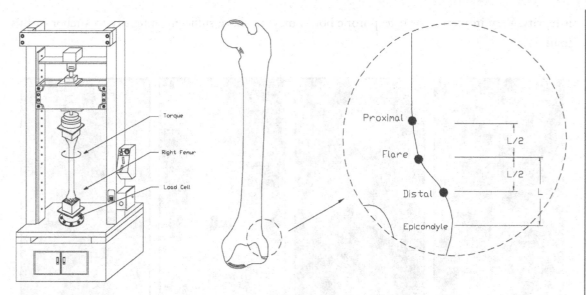

Figure 5.47: To obtain bone for fracture fixation, bone may be harvested from a patient's distal femur. A hole is drilled and the cancellous bone harvested for packing in the fracture site. Here the effect of the hole location was investigated in torsion. Again, while the bone is compromised, the question addressed was whether or not the bone was compromised to the point where fracture occurred as a result of the hole and harvesting procedure. Because torque is a function of the distance from the neutral axis and the distribution of the mass (polar moment of inertia and resistance to torque), torsion is an important test when investigating surface modifications.

Another example of a torsion application is shown in Figure 5.49. Hips, particularly in adolescent boys have the potential to shear at the femoral growth plate, a condition known as slipped capital femoral epiphysis (SCFE). Animals such as calves undergo large growth in the early years of their lives and are also subject to growth plate shear. Given this fact they are an excellent model of study. The system shown is designed to test a cow femur that has been sheared at the growth plate and stabilized with either one or two screws. The critical loads in this case are determined to be shear and torsion. The key to applying the torsion is to ensure that the torsion is applied about an axis perpendicular to the plane of the sheared physis. To accommodate this, the femur is potted in a jig that is held at an adjustable angle. The head of the femur is held to the ram/mover by a fixture containing pointed set screws. This accommodates unique geometries and a sufficient number of screws are run into the femoral head to ensure that the head can be torqued relative to the fixed base. Post-failure, a femur with the fixture holes is shown. The failure site indicates that the hold is sufficient. Had it not been, the holes would not be round, but would appear as elongated ellipses as the screws tear through the femoral head. This type of fixture works well when the bone has good integrity. Very immature or osteoporotic bones may not have sufficient integrity to anchor in this manner.

Figure 5.48: Cadavers were tested in torsion to determine the effect of the hole location, creation and harvest on the bone's structural integrity. The more distal the hole the less it compromised the bone in torsion. All proximally placed holes significantly compromised the strength of the bone. The standard torsion fixtures were utilized and the bones were potted on axis in bone cement. Care was taken to ensure that the cement volumes were consistent across the specimens, and all specimens, regardless of hole location, were potted to the same level.

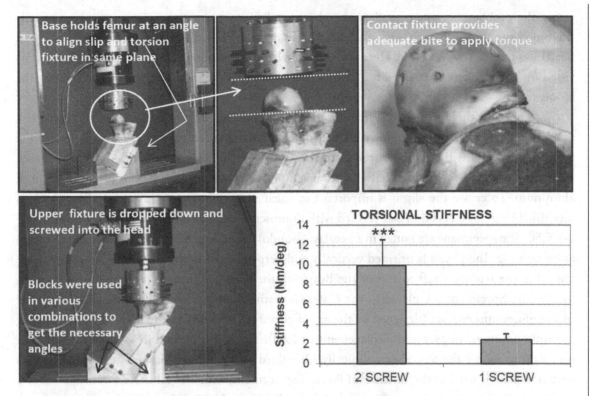

Figure 5.49: Following the creation of the SCFE model in the calf femur and fixation with either one or two screws, the bones were tested to failure in torsion. Here the cup bit into the head of the femur and twisted it relative to the fixed shaft connected to the stationary frame of the loading machine. Given specimen variability, all femurs were not torqued in this setup, but the blocks were used in various combinations to achieve torsion on the sheared plane of the instrumented SCFE.

5.2.5 SHEAR APPLICATIONS

In shear testing, one region of the object is moved relative to an adjacent region. Here we show an example of a shear test in orthopaedic biomechanics applications. We have encountered the need for shear very infrequently in our work. Shear is defined in this example not as the type of load applied, but in the load experienced by the object being tested (e.g., bone). To determine the performance of an object in pure shear, an Iosipescu test is conducted.

As briefly touched upon, in the case of slipped capital femoral epiphyses (SCFE), the femur shaft is sheared relative to its head which is fixed in the hip joint. This occurs not only in growing adolescents, but in other larger animals that grow quickly in the first years of their life, such as cows. To investigate whether one or two screws are ideal to stabilize a slip is the question the study addressed. While the answer might seem obvious that two screws will provide superior fixation to one screw, the problem is more complicated and must take into account the anatomy, and specifically the blood supply. With each screw the risk of damaging the blood supply (and killing the femoral head) is significantly increased. Here the study goal is to determine if the additional stability provided by two screws is warranted for adequate fixation given the increased risk of annihilating the blood supply.

To design the upper fixture, frozen femurs were purchased from the butcher and while still frozen the heads were measured with dial calipers and the contact geometry was obtained. The fixture contact with the femoral head is made with a 45° arc segment with a 6.35 cm radius of curvature mounted to the load cell and testing machine crosshead. The fixture is machined from aluminum. To create the slip it is important to shear uniformly across the surface. The physis is identified by gross inspection and verified with fluoroscopy using needles placed in the physis, Figure 5.50. The specimens are potted in fiberglass (Bondo) to the level of the femoral neck in a custom aluminum jig. The physis is oriented vertically and perpendicular to the custom jig. Two screws are placed in the femoral shaft to anchor the Bondo to the bone and prevent rotation during creation of the slip. Specimens are placed in the materials testing machine and the physis aligned parallel to the aluminum contact (designed so the arc of curvature is similar to the femoral head, allowing multiple points of impact and displacement of load). The slip is created in an anterior-posterior direction to mimic the forces seen clinically. One third of the average femoral head diameter (per matched pair) is used as the distance of linear displacement, with vertical shear forces applied at 1 mm/sec with load displacement recorded and the slip secured with either one or two screws.

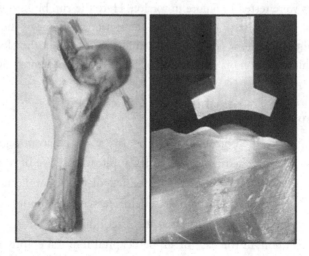

Figure 5.50: To study SCFE, SCFE was created in a calf model. The physis was located using fluoroscopy and marked with needles. The bone was potted (Bondo) and placed in the loading machine such that the growth plate was sheared (to 1/3 diameter) by contact along the femoral head with the specially designed fixture.

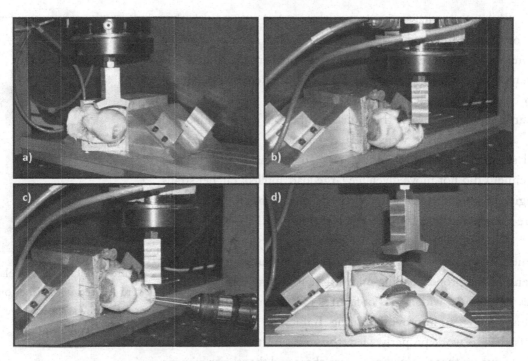

Figure 5.51: Creation of the non-reduced SCFE. Specimens were loaded to 1/3 diameter head displacement in the loading machine. Once loaded, the specimens were pinned in position and the load removed. Specimens then underwent either single or double screw fixation and then were tested to failure in either torsion or shear.

The fixturing to create the slip is illustrated in Figure 5.51. Using a cow bone (< 1 yr) a fixture was fabricated to create the SCFE by applying a compressive load to the head of the femur. Given that the diaphyseal shaft is fixed, the compressive load to the femoral head induces a shear stress on the bone shearing the bone at the growth plate. Often in biomechanics a fracture is created and the bone ends are then placed neatly back together and plated (creating a "reduced" fracture). In the clinical occurrence of SCFE, the growth plate often shears chronically over time, and the femur never completely separates. In these situations the treatment of the slip is to pin it in place to keep it from progressing and to not completely separate the bones and reduce the fracture (leaving it "non-reduced"). To model the non-reduced slip the loading machine is used to apply a load and shear the plate a fixed distance (consistent with the surgeon's clinical practice) (5.51a). Once the distance is obtained, the loading machine is stopped and the load maintained (5.51b). Pins are drilled into the femur to hold the slip in place (5.51c) and the load is removed (5.51d). The specimens are then removed from the loading machine and fixed with either one or two screws, Figure 5.52. The specimens are then randomly assigned to failure in torsion or shear. Torsion was addressed in the previous section.

A clinical concern for the surgeon in these cases is whether or not the non-reduced fracture will continue to shear or stabilize. Therefore it is important to also address further slip (shear) following the procedure. Given that alignment is altered by the initial slip creation, pure shear is not accomplished by simply using the shear creation fixture in the same orientation. To accommodate the angles, steel rods were inserted under the specimen holder and secured by the aluminum blocks. To further ensure that the specimens did not move given the large load applied, ratchet straps longitudinally and laterally secured the specimens. The specimens were then tested to catastrophic failure, Figure 5.53.

Figure 5.52: SCFE models were secured with either one or two screws, shown here with two screw fixation.

This SCFE project also demonstrates a very important take-home message in orthopaedic biomechanics. Namely, the goal is to ensure the proper loading orientation at all times. If the budget does not allow a wide series of fixtures to be fabricated, it is often just as effective to improvise fixturing. It does not have to be pretty, it just has to work properly. Here the rods and the ratchet straps help to ensure that the loading is pure shear and that the specimens do not rock or pivot during loading. Often these issues arise because the considerable loads required in some biomechanics tests are not always appreciated. In the SCFE study, failure in shear involved considerable forces that bent the screws.

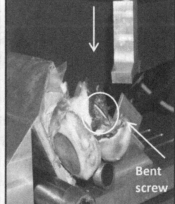

Figure 5.53: SCFE models tested to failure in shear.

5.2.6 MISCELLANEOUS HOLDERS

In addition to the fixtures discussed above, it may be necessary to hold specimens of irregular geometry. For these purposes, a generic holder that can anchor the object is beneficial. One way to hold specimens of irregular geometry is the holder shown in Figure 5.54. This hemispherical chuck has alternating rows of holes drilled and tapped to accommodate set screws machined to a point. The specimen end is placed in the holder and the screws can be tightened down as needed to engage the specimen where needed. These holders can be machined in any size and used to either apply load to a specimen via the attachment to the mover (SCFE torsion example, Figure 5.49) or used to hold a specimen to the stationary base of the loading machine.

Figure 5.54: Hemispherical chucks hold specimens of irregular geometries.

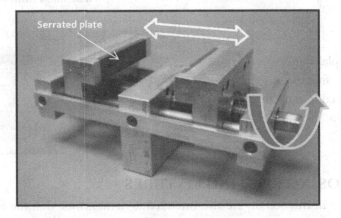

Figure 5.55: Center vise keeps specimens aligned on center at all times during testing. Commercial ones can be cost-effectively adapted to the loading machine frame.

As previously noted a useful fixture in any laboratory is a vise. The vise can be scaled up or down to accommodate a wide variety of testing needs. The vise previously discussed in Figure 5.32 was built on center to be used with the small-scale loading machine, given the adjustable base of the milling machine table. If an axial loading machine with a fixed base is used, a fixture such as that shown in Figure 5.55 is advantageous. Although the design and machining is much more intricate, this centering vise is designed using a left- and a right-handed threaded rod such that a clockwise turn will bring the vise ends closer together and a counterclockwise turn will move the ends farther apart. Using the same size threads on the left-hand and right-handed threaded rods, results in both vise ends moving the same distance with each turn. A combination wrench can be used to open and close the vise. In addition, the serrations on the vise help with gripping specimens.

Finally, when designing a fixture consider if the fixture will be needed on multiple machines. Adaptors, such as the all-thread studs shown in Figure 5.56, can be used to adapt fixtures as necessary. For example, the small-scale loading machine uses a ½-20 all-thread stud to connect the fixtures to the frame. The commercial machine in our laboratory requires #10-32 connections. In Figure 5.57 a ½-20 all-thread stud drilled and tapped for a male #10-32 stud is used to connect the three-point bending fixture to the commercial loading machine. Well designed and fabricated fixtures can work on a multitude of loading systems. Often the expense and effort spent in designing and machining quality fixtures more than pays for itself given their long-term use. These bend fixtures are routinely utilized in our laboratory and were fabricated more than 15 years ago.

Figure 5.56: A variety of connectors machined from all-thread.

Figure 5.57: The simple connectors enable the three-/four-point bend fixtures to be used in multiple loading machines.

5.2.7   REPURPOSING EXISTING FIXTURES

The figures presented in this section are also intended to demonstrate that the fixtures can be used in various combinations. For example, the femoral head contact used to create the shear slip in the SCFE study (Figures 5.50, 5.51 and 5.53) and a torsion fixture (Figures 5.45 and 5.48) were used in combination to induce bending via a compressive load on the femoral head (Figures 5.38 and 5.44). Before starting to design any fixtures it is always good practice to try to use existing fixtures. Budgets for research are often limited and time lines are often short. Developing new fixtures can unnecessarily add cost and delay to conducting a study. It is also important to look at fixtures not as the assembled components but as the unassembled pieces. Do not forget to consider using the parts in new ways. An example of this is shown below, Figure 5.58. The three-point bend fixture consists of a heavy, slotted base plate (p) to which the point contacts (c) are individually attached. The pieces are then reassembled to create the epiphyseal slip in the SCFE study as well as to then load the bones to failure in either torsion or shear. To accommodate the SCFE study, the standard three-point bend fixture is modified such that the contacts are drilled and tapped to receive aluminum brackets that attach to keep the specimen in place during torsion testing. The only new fixture fabricated for the SCFE study is the specimen holder designed after literature consultation. All other components are modified from existing fixtures. To test the slip in torsion it is critical to align the slip with the torsional plane. To accomplish this, the specimens have to be held at an angle. Given the slotted base plate, the contacts are drilled and tapped on three faces and can be manipulated to orient the specimen, as needed, Figures 5.49 and 5.58c. When repurposing fixtures, do not lose the ability to return the fixtures to their original use. If there are fixtures that are not intended to be used again, consider using the materials and components in new fixtures. Most labs will have a stock pile of materials collected over the years and metal materials (e.g., plate) can be machined for new purposes. For example, the polycarbonate and Plexiglas stock pieces used to create the tension fixtures in Figure 3.18 were leftover/scrap material found in the shop.

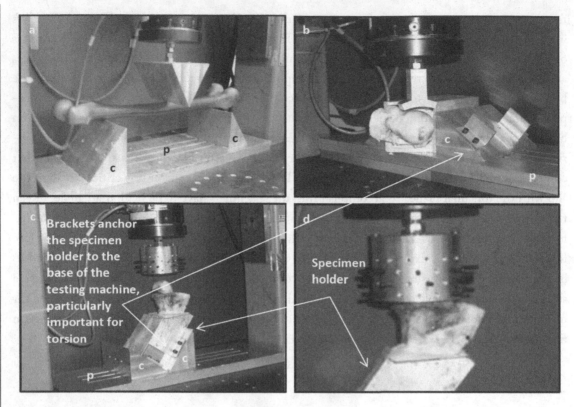

Figure 5.58: An example of repurposing fixtures. The three-point bend fixture consisting of the contacts (c) and base plate (p) are minimally modified by drilling and tapping holes to hold brackets such that the same fixture can be used in shear and tosion testing of the SCFE model. The only new fixture developed for this study is the specimen holder to accommodate the diaphyseal shaft (c–d). These components have the ability to be used in multiple studies.

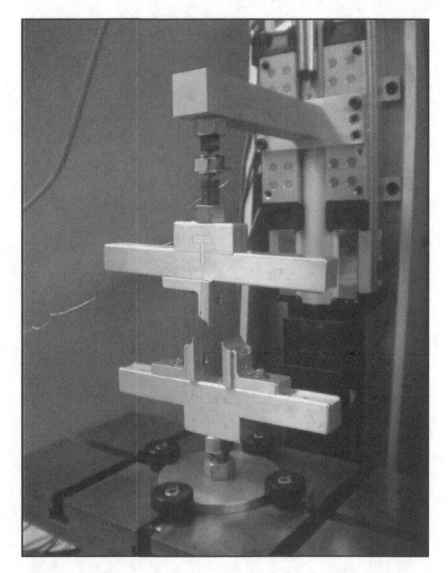

Figure 5.59: Small-scale loading platform augmented with bending fixtures. The bending fixtures are set with three contacts for three-point bending with the blade contact.

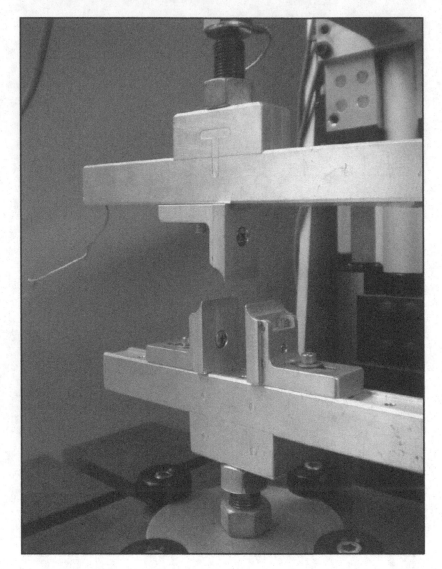

Figure 5.60: Photograph of three-point bending setup.

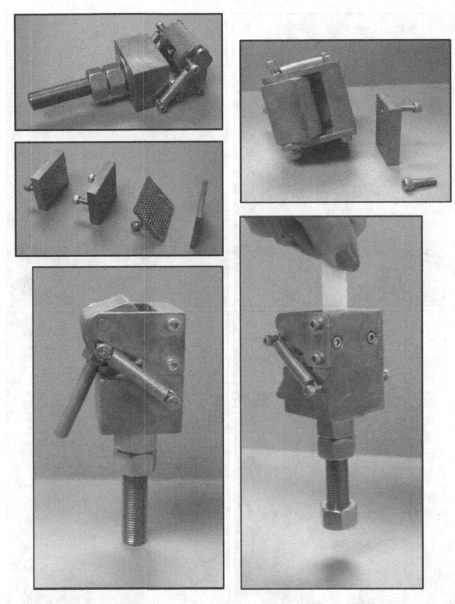

Figure 5.61: Soft tissue fixtures with knurled faces and plates of varying thickness to accommodate a range of specimen thicknesses. The springs continue to apply pressure to the specimen during testing. As shown, the clamp is strong enough to hold the paper in place.

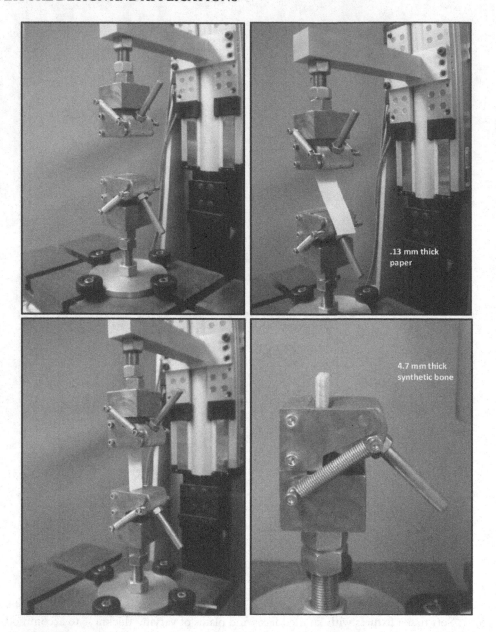

Figure 5.62: Soft tissue fixtures with knurled faces and plates of varying thickness to accommodate a range of specimen thicknesses. The springs continue to apply pressure to the specimen during testing, and the ability to rotate the handle and open the spacing for specimen insertion negates the need for wrenches when setting up tests. The setup here was for fixture illustration purposes and not actual testing. As such load cells and displacement sensors are intentionally omitted.

CHAPTER 6

Additional Considerations in a Biomechanics Test

6.1 ADDITIONAL DESIGN CONSIDERATIONS

While there is no magic formula for designing a biomechanics study, there are several issues that continuously need to be addressed. Here we address a few of them. Biomechanical engineers can find themselves working in environments with surgeons where many of the research questions will be raised by the clinician who could potentially have very detailed directives on the way in which they want a study conducted, including the study question (maybe they want to compare a fixation technique), the model they wish to use and the loading mode they want used to evaluate the fixation technique. Often these projects originate in the operating room over ways to address fractures and deficiencies in the literature that do not provide surgeons with the requisite data they want to make clinical decisions. On the other end of the continuum biomechanical engineers can find themselves dealing with a study in which they are asked to provide all the relevant details and they need to be properly prepared for such responsibility. Being aware of some of the options at their disposal can aid in sound decision making in study design.

6.1.1 KNOW THE LITERATURE

The very first thing to do when beginning a biomechanics study is to consult the literature. This is not a quick task and if a considerable amount of research has been conducted in the area, this can be a very time consuming process. But it is absolutely essential to know the current status as well as seminal articles on the research topic. Be exhaustive with search efforts, work with multiple databases, be creative in selecting search words and use both medical and engineering databases. Do not rely solely on generic Internet searches and websites. In addition to determining the research history of the topic the literature will also provide a wealth of examples on appropriate model selection, testing equipment and technique, and data analysis. Initially, a thorough literature search may take several weeks to complete but should not be rushed. The literature is an invaluable source of knowledge and with experience it will become easier to efficiently examine. Many of the issues that will be touched upon in this section may be thoroughly researched in the literature.

6.1.2   ASTM STANDARDS

The American Society for Testing and Materials (ASTM) is an organization that provides standards for a multitude of products, materials and testing applications. While the organization is not focused solely upon biomedical engineering–related standards there is a significant amount published by this society related to the biomedical engineering field. For example, there are a variety of mechanical testing standards for testing biomedical devices, as well as biomedically related materials such as elastomeric materials that may be used in biomedical applications. While it is not mandatory that the standards be used in a test, it is highly advisable. As discussed many times in this book a key goal of testing is ensuring that the data is reliable. These consistent standards help to ensure that all biomechanics testing laboratories are conducting tests in the same manner, making the results comparable across laboratories. For example if you are working in industry in a company that has developed a new artificial hip, you will want to conduct the testing according to these standards so your results are comparable to those of other hip implant designs tested in the field following this standard. If you are working in a hospital research lab and testing the hip implant, your testing may not follow the exact standards, but may consist of cadaveric fixation studies comparing the new hip to a current implant used in your institution under directives the surgeon supplies given perceived gaps in the literature to effectively determine the clinical value of the new design. Testing in this case may focus on simulating physiologic loading with the femur tested in bending and/or torsion. Similarly, the International Organization for Standardization (ISO) is the international society that works to accomplish similar tasks as the ASTM. As differences in the standards can exist, it is advisable to pull all appropriate standards from both organizations when designing biomechanical tests and to inform surgeons that these standards exist as they may strengthen the results of a given study if incorporated.

6.1.3   MODEL SELECTION

The model that you use for your study is critical to the outcome. Biomechanics models include cadaveric (human), animal and synthetic. In designing a study you should understand the research guidelines of your institution and the paperwork that is necessary to conduct the research. For instance, human research is approved by the Institutional Review Board (IRB) and animal research is approved by the Institutional Animal Care and Use Committee (IACUC). Both committees are made up of researchers and non-researchers from the institution and the community. Medical experts are on both committees with surgeons serving on the IRB and veterinarians serving on the IACUC committees. Institutional employees serve on both panels, as do religious leaders and community and student volunteers. The committees are tasked with determining that the potential benefits of the research justify the research and that the study design and protocols related to the subjects are conducted appropriately, safely and professionally at all times. Because biomechanics studies often require animal sacrifice, the IACUCs work to ensure the studies warrant animals, alternatives to animal research are not available and animal procedures performed are minimally distressing and as humane as possible.

Cadaveric Models

In the case of biomechanics work where cadaveric tissue will be utilized the research is often exempt from IRB approval, particularly where personal identifiers are not needed. Make sure to research the requirements of your institution. Sources for the cadaveric material may include private companies and foundations and medical schools. You will want to pay attention to the tissue you receive and determine if it will affect your study. Age, tissue quality and tissue preparation are among the factors that can affect your study. For example, if you have access to cadaver tissues from donors that are 75 years of age or older using this tissue source for a sports-related injury model (e.g., ACL knee reconstruction) may not be advisable. Or if the tissues from the medical school are embalmed you will need to determine the effect that that will have on your study. Furthermore, diseases may affect the tissue needed for your particular study. An example of this may be osteoporotic tissue or donors that have Ehrlers-Danlos Syndrome, a connective tissue disorder that affects musculoskeletal tissue quality. If you are obtaining tissue from an outside source (private company/foundation) you will be asked to complete paperwork specifying the parameters of your study including age, tissue preparation and tissue quality that is acceptable for your work. For example, in our studies we minimally request fresh-frozen, unembalmed femurs from female donors less than 55 years of age and male donors less than 65 years of age not afflicted with Ehlers-Danlos or Marfans Syndromes. The gender age discrepancy is a practical consequence of earlier onset of osteoporosis in women.

The decision to use a cadaver model brings up an important issue that can often occur in biomechanics research. Obtaining significant numbers of specimens may often be challenging from both availability and financial standpoints. We have made allowances and used potentially contaminated tissue. That is, we have used tissue that would otherwise be destroyed because it is not safe for allografting, or implanting into another living person. For instance, information on patient history may not be available, questionable information regarding a screening (hepatitis, HIV) may exist or the donor may be unidentified and the cause of death (overdose, gunshot wound) renders the tissue less desirable from a patient standpoint. Often with experience comes an understanding of the minimal risks associated with handling the tissue, and the increased experience with using cadavers in studies reduces the anxiety and likelihood of researchers cutting themselves, or otherwise being injured working with this tissue. However it should ultimately be your decision if you are willing to work with this tissue. When we have used tissue that has failed transplant screening acceptability we have made it a priority to increase lab safety with minimal double gloving, full gowning and full face (eye and mouth) protection. We use not just disposable scalpel blades, but disposable scalpels which negates the need to handle dirty blades and we replace them frequently.

We bleach and clean before and after and we make sure we document the study (photograph) as we go and dispose of bones as expeditiously as possible, making sure to inform others in the lab of the presence of such tissue. The latter is done to minimize injuries that may be sustained from the storage of broken bones that have sharp edges, Figure 6.1. We also dispose of all tissue following institutional policy and make sure that institutional staff coming in contact with the tissue is aware of the safety practices.

While we don't advocate the routine use of this tissue the practical limitations of obtaining healthy cadaver tissue in sufficient numbers sometimes makes this a viable option. But the researcher should in no way feel pressure to use tissue they are uncomfortable working with and they should make sure on entering a laboratory that handles human tissue that they have the prerequisite hepatitis and tetanus shots up to date. Also, depending upon the study motivation these studies are often initiated by surgeons and the residents/medical students will often do the majority of the tissue handling further lessening the risk to the biomechanical engineer.

Animal Models

Major sources of animal tissue include animal vendors and butchers. In addition, if you are working in a research setting and utilizing animal tissue that is generally discarded, you may be able to piggyback your research onto someone else's study such that in their carcass disposal, you are able to utilize the bones/tissues. This will minimize the use of animals by doubling up on projects, and institutions are

Figure 6.1: Cadaveric femur post torsion testing results in sharp bone fragments. Follow all safety guidelines when disposing of this tissue and avoid keeping this type of tissue around longer than necessary.

generally supportive of these efforts. It is also not uncommon for animal tissue for biomechanics studies to be purchased from butchers. It has been our experience that if tissue is purchased in this manner, our institutions require IACUC approval so you need to verify within your institution if you need approval to purchase specimens from a commercial butcher/market. However, depending upon the tissue it may or may not be deemed "waste" by the commercial enterprise. For example, we have been able to obtain bovine hocks (ankle joints) for free or at extreme discounts but have had to pay top dollar for intact porcine and bovine knees and femurs, as these tissues are involved in main cuts of meat sold by the vendor, Figure 6.2. If you are purchasing large volumes of tissues from a vendor you may find that they butcher on specific days of the week and you can coordinate schedules to pick up the tissue the same day it is butchered and treat it as fresh.

Synthetic Models

Synthetic models have many advantages that make them a viable alternative to cadaveric and animal sources. They are available in a wide assortment of bones and a variety of bone densities. Furthermore, for a given bone and density, they are relatively identical. While these bones are engineered to mimic the mechanical properties of bones (for simplified loading scenarios), their real benefit is in their use in evaluating surgical fixation techniques. For example, these models could be used to compare plate to rod femoral fracture fixation. Given that these bones are identical, jigs and molds can be conveniently used to create identical fractures and ensure consistency among fixation procedures. Inasmuch as these bones are synthetic and do not mimic the biological properties of the tissues such as the healing potential, they cannot be used for evaluating fracture fixation post-healing. They may only be used to evaluate the ini-

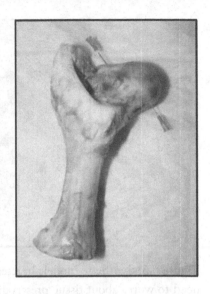

Figure 6.2: Animal models in research. Bovine limbs purchased from a local butcher provided the femurs used in a slipped capital femoral epiphysis study.

tial strength following immediate fixation. Additional advantages of using synthetic tissue include the expedience with which they may be obtained (in comparison to cadaveric tissue banks) and the lack of concern of tissue preservation, Figure 6.3. However, a significant disadvantage is that these models are synthetic and the gold standard for biomechanics is the cadaveric model. With increasing frequency, researchers will conduct a research study using synthetic bones to obtain a specimen number sufficient for statistical analysis and then will include a few cadaveric tests to verify that the trends found in the cadaver models are consistent with the synthetic results. In addition to whole bone models, cancellous bone blocks can also be purchased as a function of density with or without cortical shell (and varying thickness). These blocks are convenient for pullout studies addressing screw fixation while the varying density can enable reasonable approximation of normal to osteoporotic bone performance.

Figure 6.3: Synthetic models in research. Major advantages of synthetic models are that there is no need to worry about tissue preservation, they can be obtained in bulk relatively quickly and the identical geometry simplifies testing.

6.1.4   TISSUE CARE

If using either cadaveric or animal tissue it is important to maintain the integrity of the tissue from harvest through completion of the study. We generally store tissue (frozen wrapped in saline-soaked gauze and in air-tight bags at -20 °C) initially upon receiving it. This allows us to develop consistent testing protocols and to test the tissue under comparable environmental conditions. This is in contrast to testing one or two specimens as they arrive in the lab. The latter can introduce much more testing error via the lack of consistency. It is always preferable to have the same investigator run the tests and analyze data than to have studies drag out over a course of time such that several different people are involved. There are small biases that can arise based upon individual variations and if the same researcher conducts the studies these biases are generally inconsequential and do not affect relative comparisons.

If bones (cadaveric or animal) are to be stored for extended periods of time it is recommended to store the bones with the surrounding tissues intact. This will help to preserve the integrity of the bones and avoid freezer burn. Specimens should be completely thawed and rehydrated prior to use. Tissue care is important with respect to what is practical to assume can be tested. For instance, it is unreasonable to think that you can conduct in a day dozens and dozens of tests, even simple tests. Therefore, it is not practical to have all your tissue fresh and assume that you will complete it within that timeframe. In these instances, it is always better to freeze the tissue and thaw only what is feasible to test at one time and continue this process of treating the tissue in the same manner. Freezing and thawing multiple times has negative effects on tissues and should never be done. As a rule of thumb, if we thaw more than we can test at one time we refrigerate what is left (in saline-soaked gauze) only if we intend to test it within the next 12–18 hrs. If we cannot get to it in that timeframe we refreeze it. Be realistic with your intentions. Don't let tissue remain refrigerated for extended periods of time; this will have significant degradative effects on specimen quality.

6.1.5 EQUIPMENT

It cannot be stressed enough that it is critical to ensure that one has the appropriate testing equipment necessary to conduct testing in the laboratory. For instance, if you only have access to a uniaxial loading machine and your collaborator requires torsional loading you may not be able to conduct the necessary work. If you are willing to develop the necessary fixturing and purchase the necessary equipment (torque cells) then that is your choice. If you routinely conduct whole bone cadaveric testing you may not have the necessary equipment to complete studies on rodent limbs and vice versa. It is important to know the operational range of your equipment as well as the limitations of your systems. If your machines are not fatigue-rated you are not the laboratory to contract fatigue work from outside companies. This can prematurely wear out your equipment, as well as lead to erroneous data as testing parameters are less likely to be held accurate with the increasing wear of the parts and introduction of slop into your testing setup.

It is also important to consider the data collection capabilities. There are two rates associated with testing; the rate at which a specimen is being loaded and the rate at which the data are collected. These can be adjusted in tandem in a study setup to ensure that there is sufficient data. You want enough points such that your load-displacement/stress-strain curve is sufficient to properly interpret the shape of the curve, Figure 6.4, but not too many points that the data is noisy, Figure 6.5. When dealing with newer loading machines this issue is moot. When dealing with older systems or systems that have been built in-house such as the one we have worked with in this text, there may be physical limitations on the data collection rate that would impede, for instance, impact dynamics. Our in-house system has a maximum data collection rate of 10 Hz (10 data points collected per second). If this is too slow to adequately collect data on a fracture test we will slow down the rate of loading on the fracture test to collect the additional data. Generally, we have found that most tests we conduct can be successfully completed by altering the two rates in tandem. If this is not acceptable for the type of testing you do, for example if you need to test a significant number of ligament/tendon tissues at high strain rates, you will need to ensure that the data collection rates are sufficient to handle the testing application and a commercial system or custom loading platform for this purpose may be a better approach. While analysis of the data curve can provide a great deal of information, there can also be multiple reasons for a similar appearance in data curves. For example, if the load cell does not accurately measure in the testing range, the curve can have a similar appearance to the noisy curve shown in Figure 6.5 (right). A 10 pound (44.5 N) load cell or larger would not have the sensitivity to adequately record loads on the scale shown in the plot.

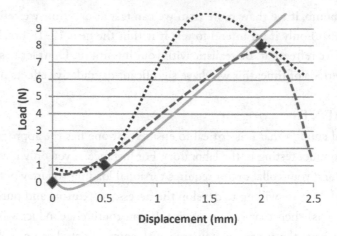

Figure 6.4: Insufficient data capture. The rate of data capture was too slow to adequately determine the specimen response. Here the three data points are superimposed with three possible curves illustrating this point.

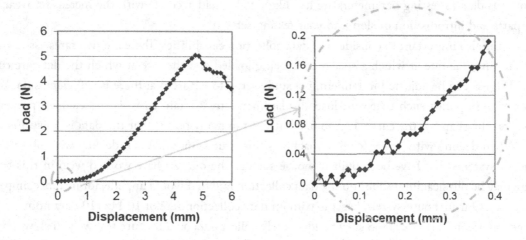

Figure 6.5: Inappropriate data collection rate. The data curve on the left was obtained from a decellularized cardiac tissue that adequately balanced the testing rate and the data collection rate and utilized load cells appropriate for the testing range. The data curve on the right was obtained from a decellularized cardiac tissue that did not adequately balance the testing rate and the data collection: the data collection rate was 6 times slower resulting in a poor signal. Be careful to interpret these curves correctly. Poor signals can also occur if the load cell is larger than the loading range and sensitivity is compromised. Testing ranges should ideally fall within the central operating range of a load cell. Although manufacturers vary, load cells are not guaranteed reliable in the extreme low and high ends of a cell's operating range. Conservatively they should be accurate within the central 90% of the range. Some companies will go so far as to claim their cells are reliable to within ±2%.

It is also important to ensure that the appropriate testing environment is maintained at all times. If specimens need to be tested in a saline bath, additional equipment will be required to maintain the environment. This includes special consideration of the fixtures as well as the sensors. Submersible load cells are available and may or may not be necessary depending upon what part of the test requires this additional constraint and environmental chamber design. It may be necessary to maintain and/or monitor temperature and humidity levels. Minimally, tissues should always be prepped and tested in a well hydrated state. For short duration tests it is often sufficient to hydrate samples prior to putting them in the testing frame and test without a environmental chamber (quick failure tests). Never let tissues "dry out" then try to rehydrate them. Maintain hydration at all times. As noted earlier in this chapter, the literature provides an invaluable source of information. For example, if environmental control is needed for a longer test (e.g., fatigue) there are countless ideas for accomplishing this in the literature.

6.1.6   SPECIMEN ATTACHMENT

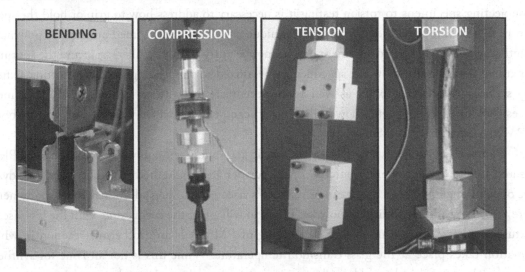

Figure 6.6: Bend testing (three- and four-point) and compression testing do not require physical end constraints during testing. In contrast, tension and torsion testing require physical end constraints applied via the loading fixture or specimen potting.

Because of the importance of fixturing in biomechanical testing, fixture design and development was considered as a separate chapter in this text. In this section we address the issue of attaching the specimen to the loading machine. As previously noted, the specimen can be thought of as the link that connects the loading machine ram (and crosshead) to the base of the machine. As such, this link is critical to the quality of the test and the accuracy of the results. The type of test to be conducted governs the means of attachment, that is, how the ends of the specimen are to be fixed or if they need to be fixed. Shown in Figure 6.6 are specimens tested in bending, compression, tension and torsion.

Specimens tested in bending or compression do not require physical end constraints to conduct testing. That is, for bend and compression tests the specimen being tested is set between the fixtures and loaded to failure, as shown in Figure 6.6. In contrast, specimens tested in tension or torsion require physical end constraints to conduct testing. That is, for tension and torsion tests the specimen ends are physically held in place, as shown in Figure 6.6. For the tension test, the specimen is clamped between two serrated friction clamps; for the torsion test the specimen is potted securely in cement such that during testing, one end of the bone is rotated around the other, which is fixed. Clamping specimens for tension testing has been previously addressed in this text. Here we briefly address some acceptable potting systems for torsion testing.

6.1.7   POTTING MEDIA

In subjecting specimens to torsion testing it is necessary to address how to grip or hold the specimen in the loading machine. A common technique is to use a potting medium such as polymethylmethacrylate (PMMA) bone cement or automotive putty (plexiglas filler) to secure the specimen in fixtures that in turn allow the specimen to be affixed to the loading machine. Potting fixtures may serve as the testing fixture as well as the receptacle for potting. When anchorage of metal pieces is required liquid metals (low boiling point) can be used, but we will focus on potting bones in PMMA and plexiglas filler.

PMMA is in general referred to as an acrylic glass and is the equivalent to commercial Plexiglas and Lucite. Bone cement is purchased as a powder and liquid methylmethacrylate. An advantage of using the PMMA is that it can be made less viscous by adding additional liquid monomer or more viscous by adding additional powder. The ratio will affect the curing time with the less viscous mixture requiring more time to cure. The curing of PMMA is an exothermic reaction in which formation takes place at the glass transition temperature. While this is not an issue for synthetic bone models, it is critical to address for native tissues since these elevated temperatures can cause thermal necrosis (death) of the tissues. In addition, PMMA is hydrophobic and difficult to bond to wet and oily surfaces. Bone marrow and fat should be removed and a degreaser (or alcohol) may be necessary to secure the bone surface to the cement. The geometry of the bone ends may be used advantageously for many of the long bones. For example, if the goal of a study is to evaluate midshaft (mid-diaphyseal) fracture fixation techniques, the distal bone ends can be completely secured in the potting medium, as shown in Figure 6.7b.

If it is not possible to pot to this level or the bone being testing does not allow for this fixation approach, a simple solution can be to secure some hardware store screws into the bone ends to give the cement something to bond to, as shown in Figure 6.7c and Figure 6.7d and Figure 6.8. In addition, given the exothermic reaction of the bone cement curing and the damage that the heat can cause, it is best to try to minimize the amount of bone cement used in relation to the volume of the bone being potted. It is important to keep the bone hydrated throughout potting and testing. Work with PMMA in well ventilated areas and read and follow all safety instructions. Mix in plastic bowls and do not allow the materials to harden in the bowl, or they will not be reusable. Never use styrofoam as the liquid monomer will dissolve it. Disposables (plastic silverware and coated waxed bowls) work well for mixing PMMA.

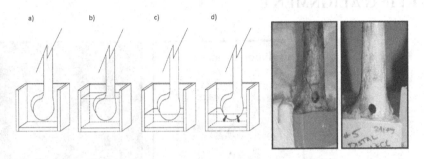

Figure 6.7: Potting is used to enable the specimen being tested to be held in the loading machine. For the femurs, potting the ends in cement blocks ensures that they can be dropped into the receptacles used to hold the bone during torsion. In addition, the potting ensures that the bone remains stationary in the bone block. To address the latter, it is critical to determine how much of the bone needs to be left exposed (unpotted) in order to accurately conduct the test. If the central bone shaft is to be tested, the bone can be potted such that the condyles are completely embedded (b). For the femur study, the photographs show that the distal bone region was tested and the condyles could not be completely embedded (c). To ensure that the bone did not pull out of the cement during testing, screws can be placed in the bone ends to provide cement anchorage (d).

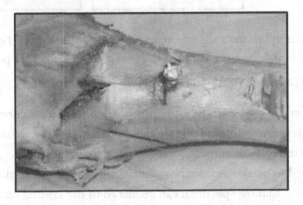

Figure 6.8: Hardware screws placed in the bone give the potting medium anchorage.

Bondo® (or auto body filler) is a great potting alternative to bone cement at a fraction of the cost. Bondo® is a resin that is mixed with a hardener to activate setting. However, this filler is highly viscous and not appropriate for potting small animal bones or situations in which low viscosity filler is needed to accommodate surface topography and crevices. Like PMMA, automotive filler will not adhere well to oily surfaces, and the hardware screws can aid in anchorage. Unlike PMMA, Bondo® will not hold up to the rigors of machining.

6.1.8 POTTING ALIGNMENT

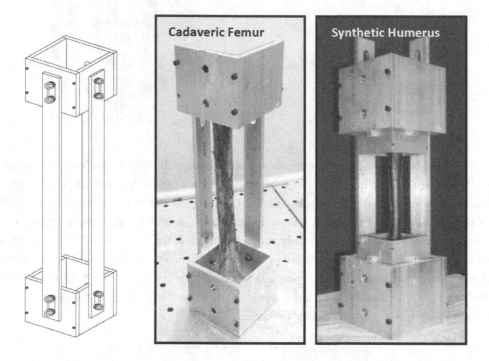

Figure 6.9: Potting fixture to maintain alignment and on-axis testing (left and center). Rather than creating multiple sets, we made inserts to use the larger alignment system for potting smaller bones (right).

When potting the bones for torsion testing, alignment is an important concern. Alignment is dictated by the physiological constraints of the model. For example, the femoral shaft has a natural curvature and may be tested with respect to an anatomic or a geometric axis. Once the desired alignment is determined, it is also necessary (particularly for torsion testing), to ensure that the bone ends are potted such that the specimen can be placed in the loading machine without any pre-load. This can be accomplished by developing a simple potting system that incorporates alignment.

An example of this is the system shown in Figure 6.9 that includes two end receptacles connected by braces. Slotting these connecting braces enables the fixtures to accommodate bones of varying lengths, as shown in the photographs, Figure 6.9. for a cadaveric adult femur and a synthetic adult humerus. If two screws are used to anchor each end of the brace to the potting fixtures, one brace is sufficient to hold the alignment. If only one screw is used to secure each end of the brace to a potting fixture a minimum of two cross braces should be used, Figure 6.10. Machining multiple fixtures will help to expedite the potting process. During potting the bone is placed in the end receptacles and the cross braces are attached. External ring clamps can be used to support the bone and maintain orientation of the specimen during potting. One end of the bone is set in the PMMA and once cured the fixture is inverted and the other end is potted. During potting the diaphyseal shaft is wrapped in a saline-soaked chuck (absorbent pad) to maintain tissue viability.

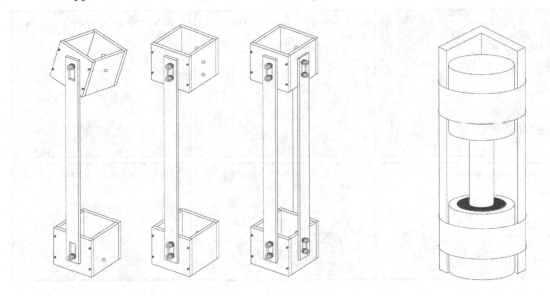

Figure 6.10: (left) Given that two points are necessary to hold plane, a single brace with one hole in each receptacle will not suffice. Either two holes per receptacle or two braces should minimally be used. We utilize two holes per receptacle and two cross braces in our potting fixtures. (right) If the budget does not allow for custom potting jigs, angle iron and pipe nipples can be used. The pipe nipples serve as the receptacles (silicon spray the inside to help with cement release) and the angle iron maintains alignment. Duct tape can be used to hold the nipples to the angle iron while the cement hardens.

When developing a potting fixture it is important to determine the extent to which the fixture will be used. For example, if conducting one torsion study, it may be unnecessary and cost-prohibitive to develop potting fixtures. If conducting a significant number of torsion studies in a variety of long bones in various models, it would be a wise investment to fabricate a potting system. However, even if needs do not justify the fixtures, it is necessary to ensure that the method of potting maintains alignment. Angle iron can be purchased at a local hardware store and cut to length. Hollow circular/square channel (or pipe nipples) can be purchased and cut to length to serve as the potting receptacles, Figure 6.10 (right). Duct taping the round channel to the angle iron will ensure that the receptacles are collinear. Rubber or metal endcaps can be used to cap the end of the receptacle and serve as the base during potting. Bone cement will not adhere to the metal surfaces; avoid internal threads in the nipples as it will impede release. Silicon spray helps to ease release of the cement from the mold after curing.

6.1.9   POTTING AND TESTING MOLDS

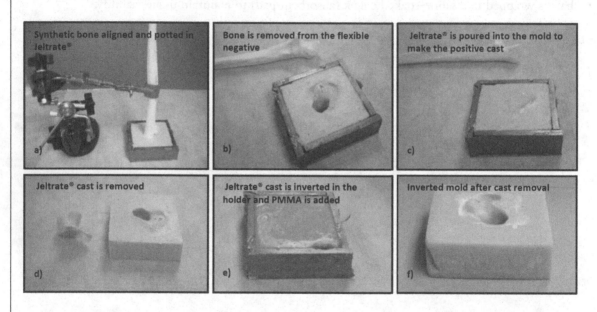

Figure 6.11: Rather than individually potting synthetic bones of identical geometry, it is convenient to make a single PMMA mold. (a) The bone is aligned in a makeshift holder and a Jeltrate® cast is made. (b) The Jeltrate® is flexible once set and the bone can be pulled out of the mold. (c–d) A positive mold of the removed region is then cast in Jeltrate®. (e) The positive is cast in PMMA and (f) removed once the PMMA sets. The PMMA is quartered on a band saw and used to hold the bones in place in the fixture during testing.

It may prove useful to make a reusable PMMA mold, Figure 6.11. This significantly minimizes potting expense and decreases time to testing. For example, we utilized a reusable PMMA mold in axial loading of a sawbone femur model, Figure 6.12. To create the mold we utilized Jeltrate®, an alignate impression material to create a negative mold of the bone end. Jeltrate® is a relatively pliable material that like PMMA is a two part mixture; the powder is mixed with water according to manufacturer's instruction. The temperature of the water determines the setting time with cold water retarding the setting time. Unlike PMMA, Jeltrate® is hydrophilic and retains good surface detail even in the presence of fluids. In addition, once set the Jeltrate® mold will not adhere to itself so it can be used to make a negative mold and then a positive casting. Jeltrate® cannot be used to make a permanent mold for testing because it is an elastic material; it is prone to tearing, and, given the water evaporation that occurs, the molds will shrink considerably as they dry out and become increasingly brittle. Because of the latter characteristic, it is advisable to make the PMMA mold from the Jeltrate® impression within a few hours of obtaining the Jeltrate® impression.

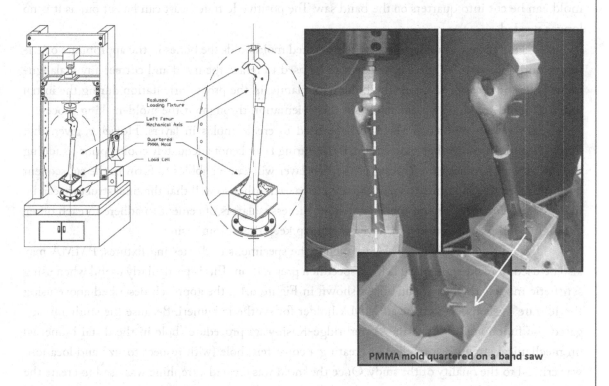

Figure 6.12: Instead of individually potting synthetic bones in bone cement, cement molds can be made for specimens of identical geometry. Here a PMMA mold was made and quartered using the approach depicted in Figure 6.11.

To make the negative mold, the bone end to be cast is placed in a container and blocked up so that it sits in the mold at the appropriate orientation. The Jeltrate® mixture is added to cover the bone end. Once set, the bone can be gently pulled out of the Jeltrate® mold and a razor blade may be used to clean up any extraneous material. If the bone does not pull out easily and tears, the mold can be held together with some plastic wrap. New Jeltrate® mixture is then poured into the mold to make a positive cast of the bone end. It is important to work with the mold within a few hours as the mold will continue to dry out and shrink. Keeping the set Jeltrate® in plastic wrap can slow this process.

The next step is to make a negative PMMA mold using the positive Jeltrate® cast. The container the mold is created in should be the fixture that will hold the potted bone during testing (or the equivalent size and shape). This will ensure that the mold will securely hold the specimen in the jig. Because cured PMMA will adhere to itself, a base layer of PMMA can be added to the bottom of the fixture. Once set, the positive cast is placed in the container (careful to maintain the final desired orientation) and PMMA is added to the desired level of the final mold. Once set the mold can be cut into quarters on the band saw. The positive Jeltrate® cast can be cut out as it is no longer needed.

It is important to make sure that the finished mold holds the bones in the appropriate orientation during testing. While a band saw can be used to reface the mold and correct any malalignment issues, it is easier to secure the bone with clamps in the proper orientation during the initial Jeltrate® mold-making step and maintain this orientation throughout the mold-making process.

As previously noted, PMMA can be used to create molds in layers. However, given that PMMA cures in an exothermic reaction, if the curing layer is not adequately cooled prior to adding the next layer of cement the heating of the first layer will cause bubbles to form in the subsequent layer of the mold. Subsequent layers will adhere to each other so well that the final structure can be machined. If for some reason you do not want successive layers of cement to adhere to each other, aluminum foil placed between the layers will help keep them from fusing.

In addition to using PMMA to anchor the specimens in the testing fixtures, PMMA may also be used to make molds that help in specimen preparation. This is particularly useful when using synthetic models that are identical. As shown in Figure 6.13, the approach described above using the Jeltrate® was used to create a PMMA holder for synthetic humeri. Because the study investigated the functional effect of the Outerbridge-Kashiwagi procedure (hole in the distal humerus) on mechanical strength, a method of creating a consistent hole (with respect to size and location) was critical to the quality of the study. Once the mold was created a trephine was used to create the hole in the mold. The specimens were subsequently placed in the mold. Duct tape was used to hold the bone to the mold during hole creation. One jig was made; the 16 mm specimens were drilled, the mold hole was enlarged to 20 mm and the 20 mm specimens were drilled.

Low melting point metal (liquid metal) can also be used to create molds or used as a potting medium. These metals can be warmed in a water bath or on a hot plate relatively safely. In Figure 6.14, liquid metal was used to make a mold to hold a hip prosthesis for mechanical testing. Shown here is one-half of the mold.

Metal machining can be used to generate molds of a high quality. An example of this is shown in Figure 6.15, in which a aluminum mold was machined according to ASTM Standard

F2118 for making bone cement specimens for testing. The two plates are clamped together and mixed bone cement is injected into the molds and allowed to harden. A little silicon spray in the molds prior to use can ease PMMA release.

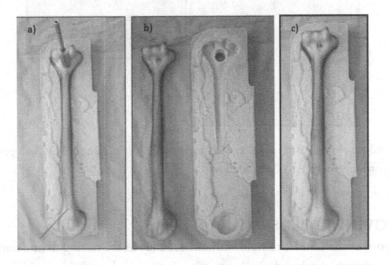

Figure 6.13: PMMA was used to make a cement jig to consistently place the hole. Once the mold was made a Steinmann pin was placed in the distal end to hold the bone while a trephine was used to place the mold hole (a), the pin and bone removed (b–c), and then used to prepare the remaining bones. Bones were duct taped to the mold, turned over and drilled.

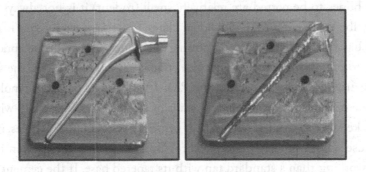

Figure 6.14: Liquid metal was used to create a mold to hold a hip plate during testing. Vaseline on the hip stem can help to ensure easy separation from the mold. Here the molds were screwed together to sandwich the implant in the fixture during testing.

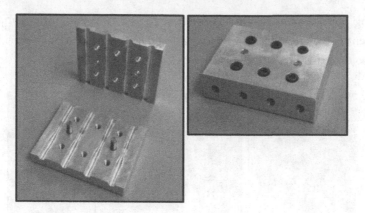

Figure 6.15: Metal machining was used here to create molds for making PMMA dogbone–shaped specimens in accordance with ASTM guidelines.

6.1.10 REMOVING MOLDS

While the purpose of the mold is to hold the bone securely in the testing fixture during mechanical testing, it is also important when designing potting fixtures and protocols to think about removing the cement from the molds. For potting larger bones (cadaver) in square or rectangular fixtures the box molds can easily be machined from individual flat plates that are then assembled to form the box. The bones are then potted in the molds and the molds can be disassembled to easily remove the test specimens. A hammer may be needed to gently tap the plate and release the cement.

When the bones to be potted are relatively small (rodent) it is not always feasible to make the molds as described above from the individual plates. If a one-piece mold is machined, drill and tap a hole in the base of the fixture. When potting the bones, use a silicone spray and place a set screw in the hole to plug it for potting. Once the bone is set in the PMMA, the specimens can be easily removed by driving the set screw into the PMMA block and out of the mold, Figure 6.16. It is critical that bone cement not get into any threads of the fixture. Cover holes with a little tape or piece of plastic to keep them clean, if necessary. If cement does get into the holes, it may be possible to drill it out or use a bottom tap. If is not a through hole, a bottom tap given its flat base has a better chance of working than a standard tap with its tapered base. If the cement can be removed, run a tap through to clean out any residual cement in the threads. If the cement cannot be removed it may be possible to redrill and tap for a larger diameter set screw.

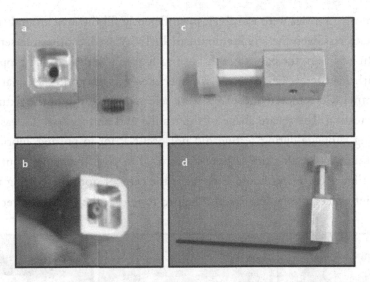

Figure 6.16: When designing fixtures, some thought should be given to the removal of the specimens from the molds after testing. If using potting receptacles that are machined from a single piece, an easy way to ensure removal is to drill and tap a hole in the base of the receptacle and plug it with a set screw. To remove the specimen, use a hex key to drive the set screw into the PMMA and push the potted specimen out of the mold.

6.1.11   SMALL-SCALE SPECIMEN PREPARATION

In addition to the rack and pinion fixtures described in the previous chapter, given the small scale of rodent work, it was necessary to develop a means by which to mount these specimens in the testing machine while ensuring that unintentional but damaging loads are not placed on them that cause premature failure. To accomplish this task, a series of potting fixtures and a testing methodology that enabled minimization of pre-test loading damage was developed. As shown in Figure 6.17, aluminum potting/testing fixtures and a slotted alignment frame with thumbscrews were fabricated that enabled the specimens to remain aligned while being potted in PMMA bone cement. The bone cement was mixed relatively thick and the specimen was lowered into the potting fixture after it was filled with cement to avoid PMMA on the shaft, which would affect the test. In addition, a square plastic template with a central hole matching the size of the specimen shaft was placed in the fixture to keep the specimen centered within the fixture. Once one end of the specimen was potted, the potting fixture was inverted and the procedure was repeated for the other end. The specimens were kept hydrated in saline, rehydrated for 60 min following potting and tested immediately thereafter. Although other methods of anchoring the bones to the fixtures were tried including cyanoacrylate (Super Glue®), nothing performed as well as the PMMA. For example, pinning the bone ends to give the cement something to anchor to was not appropriate given the size of the bones, and automotive putties were too thick to sufficiently fill the area around the specimen and ensure a rigid fixation. Given the small amount of PMMA used, exothermic setting effects were not substantial. In addition, the base of each potting fixture was drilled and tapped then plugged during testing with a set screw. Once potted, the specimen was placed in the square testing fixtures. The alignment frame was removed and plates were screwed to the top of the fixtures to secure the specimen during testing. The square shape of the testing and potting fixtures was a simple way to ensure that potting slippage during testing did not occur. Prior to potting, lubricant was sprayed in the cups and following testing the set screw was used to push the cement out of the fixture as demonstrated in Figure 6.16. This made removal of the cement from the potting fixtures relatively problem-free. Briefly freezing the fixtures also helped to release the PMMA after testing.

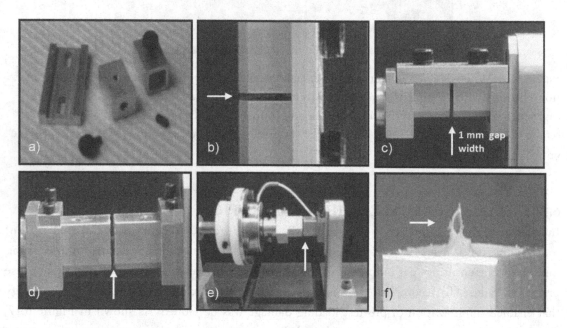

Figure 6.17: Potting system for potting bone ends in PMMA bone cement and transferring the bones to the loading machine for testing without incurring any unintentional pre-loading damage. Arrows locate exposed bone shaft.

6.1.12 MATERIAL SELECTION

Until now, little has been discussed regarding material selection. We will address some basics of material selection regarding fixture fabrication. Most biomechanics laboratories are small laboratories that have fixtures that are designed based upon the specific needs of the laboratory. Considerations that factor into this decision include frequency of use and budgetary constraints. Aluminum provides a nice compromise for material selection in biomechanics. Given financial considerations aluminum is relatively inexpensive to purchase and readily available in a wide range of shapes (rod, plate, square) and sizes. Special tooling is not required for machining and machining is thus relatively inexpensive. Aluminum is corrosion resistant, will not rust and has the strength to withstand large loads. This material may be used for bending fixtures for large mammals (including cadaveric long bone) to small neonatal rodent long bones. As such the vast majority of fixtures in our laboratory are fabricated from aluminum (6061). Because aluminum is a relatively soft material it is not appropriate for screws where thread wear would readily occur. The harder metals are more suited for screws, such as steel. However, when using steel screws it is important to consider the environment. Because steel will rust, do not use the material where it will come in contact with biological materials or need to be bleached or cleaned on a regular basis. In these cases stainless steel is a good choice. Stainless steel has excellent abrasive resistance. Stainless steel may also be used as a fixture material, but the stock material and machining costs are significantly increased relative to aluminum.

Given its ease of machining and affordability, brass may also be used in fixture fabrication. However, brass is softer than aluminum and is not ideal for larger loads. Plexiglas and polycarbonates are also used in fixture fabrication. For example, when dealing with small tissue and small scale load cells, fixture weight should be kept to a minimum. Plexiglas, polycarbonate and nylon materials work well for very small loads.

6.1.13 DATA ANALYSIS

Whether working on a commercial loading machine or one developed in-house, data analysis is a critical component of the mechanical testing procedure. Here we show a few types of data files. Minimally files will include information regarding the data collection time and transducer information. Data files generally include data point number, incremental time as determined by data collection rate, load cell readout and displacement sensor readout.

When using commercial platforms, the transducers are recognized by the system and the transducer readings (voltages) are converted automatically to units of load (load cell) and displacement (displacement sensor). Often when building your own system the transducers will need to be manually converted to the appropriate unit system. For example, the data set shown in Table 6.1 is from a test conducted on our loading system built in-house. As noted previously the scanner and displacement sensor were purchased from the same company so the displacement sensor is recognized by the scanner and the sensor can automatically be converted by the scanner software to displacement units (mm). The load cell is not recognized by the scanner and the calibration information obtained from the manufacturer at the time of purchase is used to manually convert the load cell output to units of load (N) from mV/V.

Table 6.1: Data from a test conducted on our loading system

ID	Elapsed	mV/V	mm		Load (N)	Disp (mm)
1	0.1	0	0	Slack in specimen		
2	0.2	-0.0003	0.001			
3	0.3	0	0.0267			
4	0.4	-0.0003	0.094			
5	0.5	-0.0005	0.1454	Load cell engaged		
6	0.6	-0.0003	0.1958			
7	0.7	-0.0003	0.2453			
8	0.8	-0.001	0.2947			
9	0.9	-0.0008	0.3452		0	0
10	1	-0.001	0.3937		0	0
11	1.1	-0.0013	0.4431		0.00409	0.0494
12	1.2	-0.0013	0.4936		0.00409	0.0999
13	1.3	-0.0015	0.542		0.006817	0.1483
14	1.4	-0.0018	0.5915		0.010907	0.1978
15	1.5	-0.0018	0.638		0.010907	0.2443
16	1.6	-0.0023	0.6884		0.017724	0.2947
17	1.7	-0.0023	0.7388		0.017724	0.3451
18	1.8	-0.0023	0.7903		0.017724	0.3966
19	1.9	-0.0025	0.8397		0.02045	0.446
20	2	-0.0025	0.8892		0.02045	0.4955
21	2.1	-0.0025	0.9376		0.02045	0.5439
22	2.2	-0.0027	0.9871		0.023177	0.5934
23	2.3	-0.0033	1.0366		0.031357	0.6429
24	2.4	-0.0033	1.085		0.031357	0.6913
25	2.5	-0.0035	1.1355		0.034084	0.7418
26	2.6	-0.004	1.1839		0.040901	0.7902

Converted, the data is as shown plotted in Figure 6.18. The data is from a soft tissue specimen tested to failure in tension. Here the load and displacement are plotted as absolute values and the curve initiates at the point where the load cell begins to record an actual load. In this example, the data collection and loading machine initiation are handled manually.

To capture the complete test, data collection is started before the loading machine. The delay in starting results in data points that are effectively zero. For the tension test, the test actually begins running when the displacement sensor indicates movement. While this indicates the running of the test, it is not always the first point in your data curve. For example, if there is slack in your specimen, the displacement will indicate movement but the load cell will not register a load. Therefore, in this soft tissue tensile test, the initial data point is taken as the point where the load "grabs" and continues to increase. The slower testing rate is responsible for the repeats in the load values, but these points contribute to the toe region of the curve.

The importance of understanding what data to include in testing plots is further illustrated in Table 6.2. The data in Table 6.2 represents data from a bone torsion test. The system used to collect data in this example was a commercial machine and testing initiation and data collection were synced. However, the gray region represents the slack in the test. The specimen actually began to carry load at point 29. The two curves in Figure 6.19 illustrate the importance of correctly identifying the start of

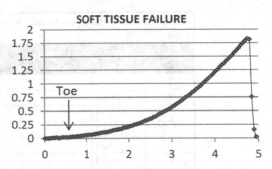

Figure 6.18: Load-displacement curve for a soft tissue tested to failure. The typical toe region and exponential response are displayed. This tissue specimen failed abruptly as indicated by the drop in load after reaching peak load.

the torque test. In Figure 6.19a the data is plotted from "point 1"; in Figure 6.19b the data is plotted from "point 29," the actual "0." Notice that the additional points incorrectly increase the failure twist and result in an erroneous value (25 degrees vs. 22 degrees). Figure 6.19a also gives the appearance of a toe region; however, hard tissues tested to failure in torsion do not exhibit a toe region. The toe region is indicative of soft tissues and fiber alignment/recruitment.

As a final example, the data curve in Table 6.3 represents a rodent long bone tested to failure in three-point bending. In this case, the change in the displacement sensor initiating at "point 1" and the corresponding "zeroes" in the load represents the time that the test was running and the dead space between the tip of the contact at the start of the test and when it made contact with the specimen, Figure 6.20. The point at which the contact initially touches the specimen signifies the start of the test. The corresponding plot for the data is illustrated in Figure 6.21.

It is critical to understand the physical significance of the data and how to plot it properly. All subsequent data analysis is contingent upon accurately plotting the data. However there is still slight variability in the data depending upon how one defines the "start" of a test. For example, in the data files shown, load is defined based upon the end of repeating numbers interpreted as load registration. Whether the last repeating number or the first non-repeating number is designated as "zero" is somewhat subjective. For this reason it is desirable, when possible, to have the same individual analyze all the data from a given test. These slight personal biases are negated if consistently applied by the same analyst. When we are beginning a study, we often collect preliminary data, plot it and determine how to define the critical parameters in the study to increase consistency in analysis. Develop an appropriate analysis strategy and then consistently apply it.

Table 6.2: Initial data points corresponding to torque-twist data plotted in Figure 6.19

Point	Time sec	Torque Nm	RVDT deg		Torque (Nm)	Twist (deg)
1	0.05	0	42.9212			
2	0.1	0	42.9212			
3	0.15	0	42.9212			
4	0.2	0	42.8968			
5	0.25	-0.195	42.848			
6	0.3	-0.195	42.7503			
7	0.35	-0.195	42.6771			
8	0.4	-0.195	42.5794			
9	0.45	-0.195	42.4818			
10	0.5	-0.195	42.3597			
11	0.55	-0.195	42.2376			
12	0.6	-0.195	42.0911			
13	0.65	-0.391	41.9691			
14	0.7	-0.195	41.8226			
15	0.75	-0.195	41.6761			
16	0.8	-0.195	41.5296			
17	0.85	-0.391	41.3831			
18	0.9	-0.391	41.2366			
19	0.95	-0.391	41.0657			
20	1	-0.391	40.9436			
21	1.05	-0.195	40.7483			
22	1.1	-0.391	40.6262			
23	1.15	-0.391	40.4553		Load cell	
24	1.2	-0.391	40.3088		engaged	
25	1.25	-0.391	40.1624			
26	1.3	-0.391	40.0159			
27	1.35	-0.391	39.8206			
28	1.4	-0.391	39.6985		Torque (Nm)	Twist (deg)
29	1.45	-0.391	39.5276		0	0
30	1.5	-1.172	39.4055		0.781	0.1221
31	1.55	-1.953	39.2102		1.562	0.3174
32	1.6	-2.734	39.0637		2.343	0.4639
33	1.65	-3.711	38.9172		3.32	0.6104
34	1.7	-4.688	38.7463		4.297	0.7813
35	1.75	-5.86	38.6242		5.469	0.9034
36	1.8	-6.836	38.4533		6.445	1.0743
37	1.85	-7.813	38.3068		7.422	1.2208
38	1.9	-8.985	38.1359		8.594	1.3917
39	1.95	-9.961	37.9894		9.57	1.5382
40	2	-7.031	37.843		10.742	1.6846

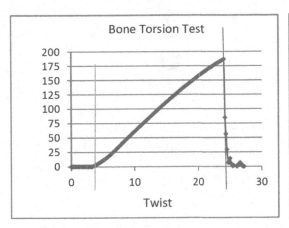
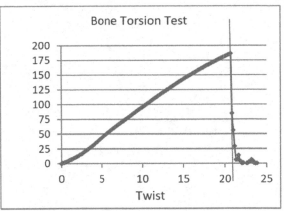

Figure 6.19: (a) Incorrectly plotted data gives the appearance of a toe region for this cadaveric femur tested to failure in torsion. (b) Correctly plotted data shows that the incorrectly plotted data results in an incorrect, excessive failure twist.

ID	Elapsed	[102] mV/V	[101] mm	Load (N)	Disp (mm)
			Table 6.3: Initial data points corresponding to load-displacement data plotted in Figures 6.20 and 6.21		
1	0.1	0	-1.8199	Load cell engaged	
2	0.2	0.001	-1.8278		
3	0.3	0.001	-1.8436		
4	0.4	0.001	-1.8516		
5	0.5	0.002	-1.8634	Load (N)	Disp (mm)
6	0.6	0.002	-1.8753	0	0
7	0.7	0.006	-1.8872	0.048955	0.0119
8	0.8	0.011	-1.899	0.110149	0.0237
9	0.9	0.016	-1.9069	0.171342	0.0316
10	1	0.018	-1.9228	0.19582	0.0475
11	1.1	0.019	-1.9307	0.208058	0.0554
12	1.2	0.02	-1.9465	0.220297	0.0712
13	1.3	0.021	-1.9584	0.232536	0.0831
14	1.4	0.022	-1.9663	0.244774	0.091
15	1.5	0.023	-1.9782	0.257013	0.1029
16	1.6	0.026	-1.9979	0.293729	0.1226
17	1.7	0.031	-2.0019	0.354923	0.1266
18	1.8	0.034	-2.0138	0.391639	0.1385
19	1.9	0.039	-2.0256	0.452833	0.1503
20	2	0.045	-2.0335	0.526265	0.1582
21	2.1	0.054	-2.0454	0.636414	0.1701
22	2.2	0.063	-2.0573	0.746562	0.182
23	2.3	0.073	-2.0692	0.868949	0.1939

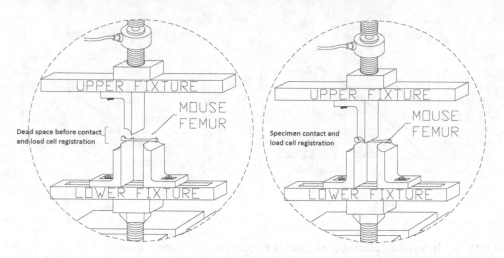

Figure 6.20: In a bending test, data analysis must account for the testing scenario. Here, the dead space is represented by movement of the displacement transducer with no increase in load. The increase in load occurs on contact with the specimen and is the actual initiation of loading.

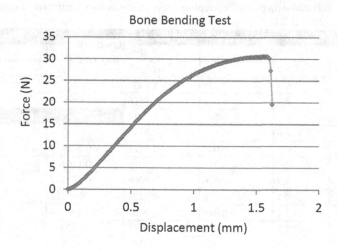

Figure 6.21: Load-displacement curve for a rodent long bone tested to failure in three-point bending.

As noted previously, the job of the biomechanical engineer is to develop sound testing protocols in order to obtain reliable data. The data are then used as the basis upon which to make the conclusions of the study and any subsequent recommendations. Remember that your work may be used by a surgeon to determine which fracture fixation system to use in a patient, it may be used by a company to get FDA approval for an implant, or it may be used to determine the effects of an osteoporotic treatment improving bone strength, etc. It is imperative to strive for excellence at all times.

CHAPTER 7

Laboratory Examples and Additional Equations

The data to complete these assignments is provided and can be downloaded: Suture data for Assignment 1; bending data for Assignment 2; torsion data for Assignment 3.

7.1 ASSIGNMENT

You work for a materials testing firm that has been contracted by a suture development company. They have four new grades of silk suture that they believe will vary with respect to their mechanical strength and as a result, will have a wide variety of uses in the medical field. Your task is to break up into two groups and conduct "blind" studies to determine the mechanical strength of two of the suture grades (that will be given to you). You will be given three strands of each grade and utilizing the materials testing machine, you will determine the stiffness, breaking force and breaking length of each of the strands under tension. For each of the suture grades you are contracted to determine an average set of mechanical properties. You will then meet with the other group and compare results.

To complete this work, you will need to know how to read a dial caliper. You will be instructed on this technique and to verify that you understand this process, please complete the following example before continuing.

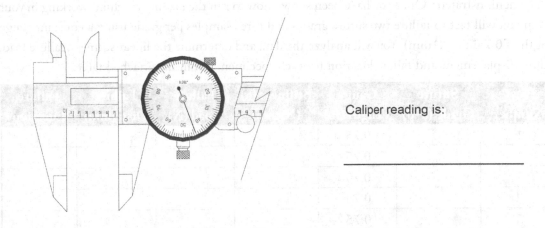

Caliper reading is:

_____

A loading platform was developed from a simple milling machine base table and a linear slide operated by a programmable controller (shown below). Affixed between the stationary platform and slide are polycarbonate friction clamps. The upper clamp is attached to a small universal joint that allows the specimen being tested to self-align. To test the sutures, a program has been written to control for a linear ramp loading waveform and a physiologic rate of loading (shown below). Data collection via Strain Smart software occurs at a rate of 10Hz. To record load-displacement data, a load cell and linear displacement sensor will be connected. The load cell records loads up to 10 lb (44.5N) and the displacement sensor has a recordable travel of 1in (25mm).

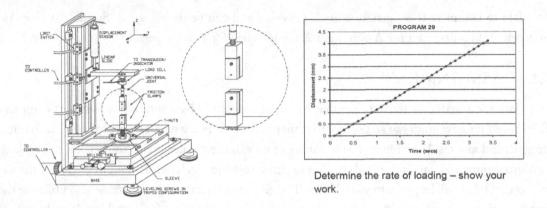

Determine the rate of loading – show your work.

7.1.1   PART I

Working within your group and with your supervisor, set up the loading machine, as shown. Your supervisor will demonstrate the proper way to set up, conduct and analyze the suture tension test. It is important to not "overload" the load cell by incorrectly tightening the upper clamp. In every case, always secure the suture in the upper clamp before securing the suture in the bottom clamp (this will be demonstrated). Once you have been shown how to run the loading machine, working in your group you will test to failure two suture grades and three samples per grade using a constant gauge length of 0.75in (19.1mm). You will analyze the data and determine the linear stiffness, failure load, failure displacement and failure location for each specimen. Fill in the attached table.

Test Number	Suture Grade (a–d)	Gauge Length (in)	Stiffness (k) (N/mm)	Failure Load (N)	Failure Disp (mm)	Failure Location
1		0.75				
2		0.75				
3		0.75				
4		0.75				
5		0.75				
6		0.75				

And the computed averages are:

Set Number	Suture Grade (a–d)	Gauge Length (in)	Stiffness (k) (N/mm)	Failure Load (N)	Failure Disp (mm)	Failure Location
1		0.75				
2		0.75				

Understanding that sutures are like wire, that is, the smaller the number the larger the diameter (i.e., 5-0 (00000) is smaller than 1-0 (0)), get together with the other group and determine the results of the "blinded" study, by "unblinding" the results. That is, determine what grades were tested.

Grade	a,b,c,d
1-0 (0)	
2-0 (00)	
3-0 (000)	
5-0 (00000)	

Do the results make sense from a practical standpoint? Explain your answer.

7.1.2 PART II

As you know, sutures are held in place by tying knots. The company also wants to know if a knotted suture will be mechanically different from an unknotted suture. Your next assignment is to break two strands of knotted suture given to you by your supervisor. Follow the exact protocol as before and determine the same average mechanical properties.

Test Number	Suture Grade	Gauge Length (in)	Stiffness (k) (N/mm)	Failure Load (N)	Failure Disp (mm)	Failure Location
1		0.75				
2		0.75				
Average		0.75				

Did knotting significantly affect the suture's mechanical properties? Explain your answer.

Realizing that the knotted sample is 3-0 (000), determine the average decrease in failure load as a function of the knot. Show your work.

Did knotting the suture affect the failure location? Explain your answer.

Based upon your answer, are the grips provided appropriate for conducting suture testing? If not, how would you modify them (what are the issues of how it fails)?

Helpful Conversions and Calibration Values:
   Load Cell: 10 lb / 2.103 mV/V
   25.4 mm = 1 in
   4.45 N = 1 lb
   Slope = change in rise / change in run

7.2   ASSIGNMENT

You work for a materials testing firm that has been contracted by a research group studying bone biology. They have long bones from genetically modified adult mice that they want to characterize in three-point bending. They have submitted several unsuccessful grant proposals and repeatedly get criticized for the lack of biomechanical data in their grant applications. They paid a consulting company to do the work and although the company completed the testing, they are no longer in business and can provide no additional information. The biologists have contacted your firm and have asked you to re-evaluate their data. They have asked you to obtain as much information about the bones as you can, although they have little input into what this data is.

You and your colleagues have obtained eight data sets from three-point bend tests. For reasons unknown to you, four of the bones were tested with one fixed distance between the lower contacts (gauge length) (1.5 in) and four of the bones were tested with another fixed distance (0.75

in). They have obviously used a dial caliper in English units as the data for bone diameter are also provided in inches. You need to report your final data in metric. Analyze the data for both structural and material properties and determine if the change in the distance between lower contacts has an effect on the data, or if it can be accounted for and the data may be used as one set (n=8).

Because you are a co-op student, a senior engineer gives you the project and some information to help get you started. He/She shows you how to read a dial caliper and suggests that you start analyzing the data by determining bending stiffness, failure load (and moment) and failure displacement. He/She also suggests that you can assume a round, hollow cross section (with inner diameter ½ that of the outer diameter) to determine the material properties, including bending stress, inertia and modulus for all eight bones (keeping in mind the contact distance difference). Plot the data and give a final report including tabularized data and a PRACTICAL EXPLANATION OF YOUR DATA that shows you can communicate the ENGINEERING concepts to a NON-ENGINEERING audience. In an attached appendix, show all mathematical work to justify your findings.

Any time measurements are collected, it is important to take a minimum of three readings of the same location and average the results. The distance between the contacts, as well as the bone dimensions were all collected this way. The final dimensions you use will be the average of three measurements.

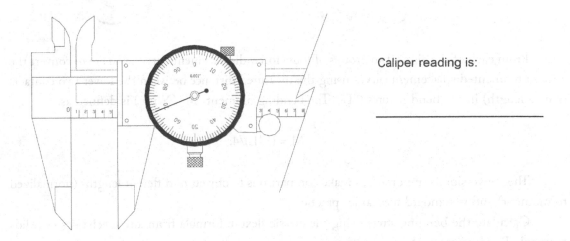

Caliper reading is:

_____

The loading platform used to obtain the data is shown in the Figure. It was developed from a simple milling machine base table and a linear slide operated by a programmable controller. Affixed between the stationary platform and slide are aluminum three-point bend fixtures. To test the bones, a program was written to control for a linear ramp loading waveform and a physiologic rate of loading. Data collection via Strain Smart software occurs at a rate of 10Hz. To record load-displacement data, a load cell and linear displacement sensor were connected. The load cell records load up to 10 lb (44.5N) and the displacement sensor has a recordable travel of 1in (25mm). Although these transducers are purchased and designated in English units, it is standard research practice to convert all data to metric. MAKE SURE YOU ARE CORRECTLY WORKING WITH METRIC UNITS AT ALL TIMES!!

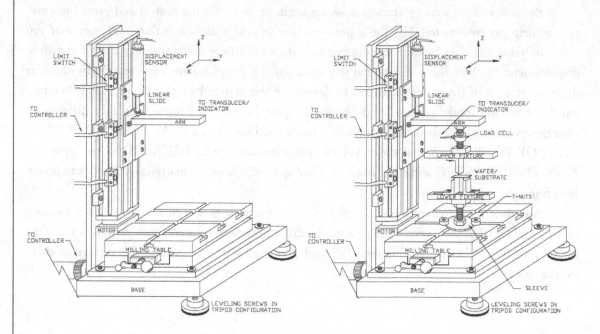

From each of the eight (two groups of four) force-displacement curves, you are to convert the data to moment-displacement curves using the measured distance between the lower two contacts (gauge length) in the bend fixtures ("L"). The bending moment "M" (Nmm) is defined as:

$$M = (F*L)/4.$$

The conversion is important to make comparisons to objects of different lengths (normalized to moment) and is standard mechanics practice.

Calculate the bending stress using the classic flexure formula from any mechanics of solids textbook. Bending stress "S" is defined as:

$$S = (M*c)/I,$$

where "S" is the bending stress (N/mm2), "M" is the bending moment, "c" is the distance from the farthest point in the cross section to the neutral axis (you are assuming ROUND cross sections) and "I" is the area moment of inertia. Assuming a prismatic beam, the area moment of inertia may be calculated for a hollow circle as:

$$I = \pi/64 * (X_{max} * Y_{max}^3 - X_{min} * Y_{min}^3).$$

Stress allows comparisons to be made between strengths of bones that differ in length, size and shape. For three-point bending, breaking stress is calculated as:

$$BS_{max} = M_{max}*c/I = ((FL*L*c)/(4*I)),$$

where "BS" is the breaking strength (N/mm2), "FL" is load at failure,"L" is the distance between end supports and "c" (again) is the distance from the centroid to the surface and "I" = area moment of inertia.

The modulus of elasticity for bone in three-point bending is defined as:

$$E = (F^* * L^3)/(48* I* d_a),$$

where the previous definitions hold and "da" is the deformation at the point of load application, measured as actuator displacement. When selecting an appropriate "F," remember the region in which "E" is defined.

Finally strain "ε" takes into account the deformation (bending) that occurs in the bone as it undergoes loading. It is a unitless parameter representing a change in length per unit length. In three-point bending, this is calculated as:

$$\varepsilon = (12 * da * c)/L^2,$$

where the previous definitions hold.

Helpful Conversions and Calibration Values:
   Load Cell: 10 lb / 2.103 mV/V
   25.4 mm = 1 in
   4.45 N = 1 lb

You are to complete the tables below.

Specimen	Diameter (in)	Average Diameter (mm)	Length (in) (L)	Length (mm) (L)
1	0.099; 0.099; 0.098		1.5	
2	0.098; 0.097; 0.097		1.5	
3	0.099; 0.098; 0.097		1.5	
4	0.097; 0.099; 0.099		1.5	
5	0.098; 0.098; 0.096		0.75	
6	0.099; 0.097; 0.096		0.75	
7	0.096; 0.096; 0.098		0.75	
8	0.098; 0.098; 0.098		0.75	

Specimen	Fail Disp (mm)	Fail Load (N)	Moment Max (Nmm)	da (mm)	c (mm)	I (mm^4)	BS Max (N/mm^2)	E (N/mm^2)	Strain
1									
2									
3									
4									
5									
6									
7									
8									

In 1-2 paragraphs, explain the data in layman's terms and if they make sense. In addition, in 1-2 paragraphs, determine if the difference in contact points (L) has an effect on the data, how you arrived at this and if it makes sense from an engineering perspective. Show that you have thought about the data. Again, this is not a group project.

You may consult references, other textbooks, etc. Cite any references that you use. You have been provided with the equations to get you started. If you do not understand them, consult additional reading from solid mechanics textbooks.

Three-point and four-point bending tests are common tests associated with long bone mechanical loading, such as the femur, humerus, tibia and radius. There are a few considerations for setting up a bend test. The first is to address whether the bend test should be three- or four-point bending. When deciding upon the type of test to conduct, the decision should be driven by the type of data that is needed. For example, in the schematic below in the top row (a) a beam is subjected to three-point bending on the left and four-point bending on the right. As shown in the second row of figures (b), the free body diagrams for three- and four-point bending are provided. In the third row of figures (c), the moment diagrams demonstrate that for the three-point bending load the maximum moment occurs at the loading point, or at

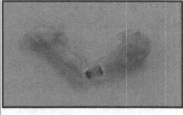

LOW ENERGY, TRANSVERSE FRACTURE

To see a video of this detail click here.

½ the span length of the lower two contacts. For the four-point bending, the maximum moment is ½ that of the three-point bending case and is constant for the span length of the lower two contacts. In the fourth row of figures (d), the shear diagrams demonstrate that throughout the length of the beam, the beam is continuously subjected to a combination of bending and shear. In the

case of four-point bending, the beam experiences pure bending between the span of the inner two contacts. As such, whether a pure bending (four-point) or combination bending and shear (three-point) is needed, will determine the type of test conducted. If a relative comparison of bending strength is needed, often practicality will dictate and three-point bending will be chosen because it is easier to ensure all three loading points will be in contact with the beam. This is particularly true if the specimens to be tested are not prismatic (such as bone) and/or very small (such as neonatal rodent bones). It will be up to the engineer to select the type of bend test conducted based upon the information needed, the available equipment and the specimens to be tested.

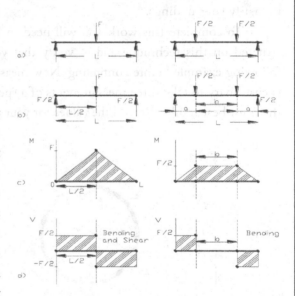

7.3 ASSIGNMENT

You work for a materials testing firm that has been contracted by a research group studying bone biology. They have long bones from genetically modified adult mice and they want to quantify the torsional properties of the bones. They have submitted several unsuccessful grant proposals and repeatedly get criticized for the lack of biomechanical data in their grant applications. Unfortunately they are not at an institution that has an engineering program and they are not able to find local engineering collaborators. Thus, they have approached your group and have asked you to conduct some biomechanical testing of these bones. The reviewers have suggested they conduct torsional studies. The biologists have contracted with your company to conduct torsional testing on bone and obtain as much information about the bones as you can, although they have little input into what is needed.

You and your colleagues have developed a protocol for loading and analyzing the data and now are ready to conduct the tests. You are to follow the protocols and determine both the structural and material properties of the specimens. You are to test four bones in torsion and making note of the gauge length, you are to determine torsional stiffness, torsional failure load and failure twist. In addition, assume a hollow, round cross section (with inner diameter ½ that of the outer diameter) determine the mechanical properties, including shear stress, polar moment of inertia, and shear modulus for a hollow cross section. Write a detailed report to the client on the RESULTS AND DISCUSSION OF THE DATA and in an attached appendix, show all mathematical work to justify your findings.

To complete this work, you will need to know how to read a dial caliper. You will be instructed on this technique and to verify that you understand this process, please complete the following example before continuing. Now measure the bone diameters and the gauge lengths. It is customary to take three measurements of a specimen diameter in the same location and average them for best results. Record the values for your calculations.

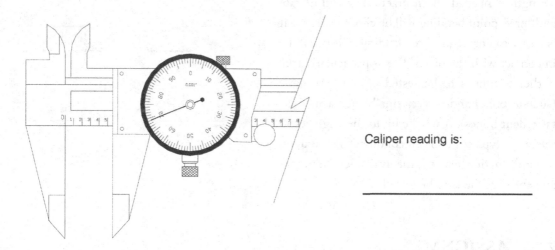

Caliper reading is:

_____

A loading platform was developed from a simple milling machine base table and a linear slide operated by a programmable controller (shown below). To enable torsion testing with the linear motion, a rack and pinion system has been developed. To test the bones, a program has been written to control for a linear ramp loading waveform and a physiologic rate of loading (shown below). Data collection via Strain Smart software occurs at a rate of 10Hz. To record torque-twist, a torque cell and linear displacement sensor will be connected. The torque cell records torsion up to 25 oz-in (176.5 N mm) and the displacement sensor has a recordable travel of 1 in (25 mm). Although these transducers are purchased and designated in English units, it is standard research practice to convert all data to metric. MAKE SURE YOU ARE CORRECTLY WORKING WITH METRIC UNITS AT ALL TIMES!!

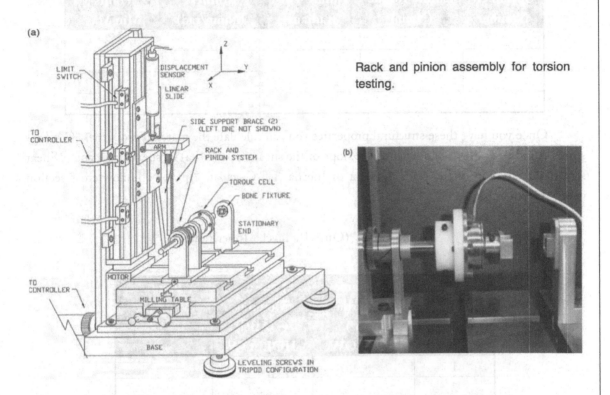

Rack and pinion assembly for torsion testing.

Record your "average" values for the bones with the hollow cross section:

Specimen Number	Shaft Diameter (mm)	Shaft Radius (mm)	Gauge Length (mm)
1			
2			
3			
4			

You have been supplied with four data curves that have been converted for you such that you have torque and twist columnar data already converted with column "E" representing torque (Nmm) and "J" representing rotation (radians). Plot your torque-twist curves.

If failure torque is defined as peak torque (Nmm) and failure twist (radians) as the corresponding angle, record the values. Calculate the torsional stiffness as the slope of the torque-twist curve in the linear region. Also record the minimum and maximum data point numbers that you used to define the linear region for your stiffness calculations.

Specimen Number	Failure Torque (Nmm)	Failure Twist (radians)	Stiffness (Nmm/rad)	Points Min Max
1				
2				
3				
4				

Once you have these structural properties you can calculate the material properties "G" (Torsional Rigidity or Shear Modulus) as the slope of the shear stress (MPa) vs. shear strain curve. Shear stress is a function of the polar moment of Inertia "J_p." Calculate "Jp" for a hollow cross section using the equation:

$$Jp = \pi/2 * (OuterRadius4—InnerRadius4).$$

Specimen Number	Polar Moment of Inertia (mm^4) HOLLOW	Shear Modulus G (MPa) HOLLOW	Points Min Max
1			
2			
3			
4			

Now you can add the column data using the equations.

Shear stress (MPa) (ζ) is

$$\zeta = Torque (Nmm)* \ radius (mm)/ J_p (mm^4),$$

and shear strain (radians) (γ)

$$\gamma = radius (mm) * twist (radians) / gauge \ length (mm).$$

After plotting the shear stress vs. shear strain curves, calculate "G" (shear modulus) in the linear region.

Compare your findings and show all calculations.

Helpful Conversions and Calibration Values:
  Torque Cell: 25 in oz = 2.08 mV/V
  Keypad (linear displacement): 1.000 units = 0.20 in
  25.4 mm linear displacement = 83 degrees; 1.45 radians
  Slope = change in rise / change in run

The assumption of a round cross section in the mid diaphyseal shaft is reasonable. Shown in the figure is a slice of rodent bone potted in bone cement and sectioned transversely on a diamond saw. A dissection scope can be used to take a photograph that can then be adequately measured using image analysis. A rule of some type is necessary to establish scale for obtaining measurements. Final dimensions are averaged from three repeated measurements.

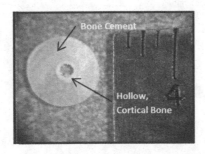

CHAPTER 8

Appendices
Practical Orthopaedic Biomechanics
Problems

In this chapter, we work to put into practice what has been applied in the previous chapters of this book. The goal of publishing a research study is to provide the interested/invested reader with information gained upon completion of an experimental study. The key is that information must be relevant and novel to the field. It must add a piece of information that was not previously known (novel) but should be known (relevant). In biomechanics, this can range from designing a new implant or evaluating a new fixation technique to repeating a study in a more appropriate animal model than what has been conducted to date in the field. There are all sorts of valid motivations for conducting a research study. The focus of the paper is on the question asked and the relevance of that question. Publications do not often focus on the details of the study but assume that the researchers know how to do their jobs and that the study findings are reliable. As such, much of the details we have tried to focus on in the book to obtain reliable data through proper use of engineering principles and design are omitted and not even discussed in the literature. But they should always be followed to the best of the researchers' abilities, as the research group is accountable for their research findings and their interpretation of those findings. Whether or not the reader takes the same interpretation is not of concern. For example, some surgeons reading a paper based upon a comparison of fracture fixation techniques may be persuaded to change their surgical treatment based upon the study, while others may not.

Here we address a few of the typical types of research studies that are found in biomechanics journals. Instead of focusing on the question, we focus on the biomechanical process which is often absent from the final publication. We will try to provide some behind-the-scenes-insight on the process of conducting a research study to help the new biomechanical engineer to make the transition from the research process to the research publication.

8.1 EXAMPLE 1: IMPLANT DESIGN: PROTOTYPE, BENCHTOP ANALYSIS

Kevin Black, M.D.
Marnie M. Saunders, Ph.D.

8.1.1 THE PROBLEM

The anterior cruciate ligament (ACL), along with the posterior cruciate ligament (PCL) and medial collateral (MCL) and lateral collateral (LCL) ligaments, make up the four main stabilizing structures of the knee joint. The ACL inserts into the femur crossing anteriorly to the PCL and inserting into the tibia. It consists of a tough fibrous material that functions to limit excessive joint mobility. While tears are not uncommon they may not be noticeable during low impact activities such as walking, but significant instability (sliding of femur over tibia) may be prevalent at higher level activities. Untreated ACL tears may result in further injury to the knee.

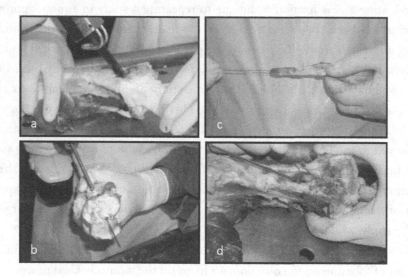

Figure 8.1: ACL reconstruction surgery using a native graft replacement held in place with Kurosaka screws. The technique is shown on a porcine knee.

A common surgical treatment of a torn ACL is to insert a replacement tissue, or graft in the location of the damaged ACL, Figure 8.1. A common graft source is the patient's own knee. A bone-patellar tendon-bone graft isolated from a portion of the patellar bone, central 1/3 of the patellar tendon and tibial bone block (8.1a), is inserted into a tunnel that has been surgically drilled through the joint (8.1b) to coincide with the orientation of the ACL. To assist in feeding the graft into the tunnel a small drill bit is used to run a through hole into each bone block and suture is run through the holes to make a "handle" on each end of the graft (8.1c). The graft is fed into the tunnel, and while placing tension on the graft, Kurosaka screws are inserted (8.1d). One screw secures the bone block in the tunnel on the femoral side and the other secures the bone block of the graft to the tunnel on the tibial side.

While the Kurosaka screw is mechanically adequate with respect to holding strength, malalignment issues are not uncommon. Visibility is hampered by the scale of the situation in which the tunnel diameter is on the order of 10mm; the bone blocks are on the order of 9mm; and, the screw diameter is on the order of 7–9mm. During placement the harder, metal screw collapses the softer bone block and ideally initiates integration of the bone, Figure 8.2. Given the relative sizes of the tunnel diameter, effective graft bone block diameter and screw diameter, the screw may not run straight in the tunnel and may result in impingement of the graft against the screw, damaging the tendon graft.

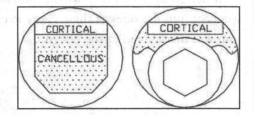

Figure 8.2: Kurosaka screws anchor graft in tunnel (planform view).

8.1.2 GOAL

To reduce placement issues associated with poor visibility by developing a technique or system to place ACL grafts under improved visibility conditions while maintaining the minimal stability of the current fixation standard, the Kurosaka screw.

8.1.3 SOLUTION

In design it is a good rule of thumb to follow the KISS adage (Keep It Super Simple) and to not reinvent the wheel. Generally, if you need a simple fastener or holding device an acceptable solution probably already exists. While it may need modifying for a biomedical application and the environment encountered in the body the basic mechanical principles remain unchanged. Here the approach undertaken was to develop a device that may be placed in a "collapsed" state to increase visibility during placement, but can then "expand" to fill the space and hold the graft as securely as the standard Kurosaka screw.

8.1.4 METHOD

The concept of an expandable implant is not that different from standard cement/concrete anchors, and a few were purchased from the local hardware store to assess design issues. Using these systems for inspiration, an expansion anchor was designed. To begin the design process computer drawing software was used to develop prototype designs. As shown in Figure 8.3, a 2D section view of the anchor in a facsimile of a bone block with the tunnel wall drilled was modeled. By trial and error, the anchor (anchor 1) was designed to expand and bite into the surrounding bone block. For these initial studies the graft was not modeled; the goal was simply to determine a design style that would meet the intended needs. Given that 7–9mm Kurosaka screws have been shown to have little difference in effect on stability, the smaller size screw (7.3 mm) was the target size. Thus, for the anchor to be successful it would have to insert smaller than the 7.3 mm. For this reason the design process was started with modeling a bone block in cross section with a 7 mm hole drilled. The anchor was then designed to fit easily within the block. To accommodate the expansion, four 0.38 mm slits were cut into the sides of the anchors to create four expandable arms. A wall thickness of at least 1 mm was maintained in the design.

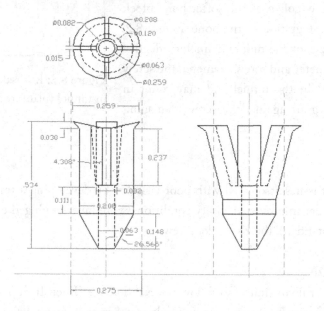

Figure 8.3: Schematic of expansion concept using side slits to expand the anchor increasing visibility during placement. Drawing is in English units per machinist's preference.

Once the initial design was determined the implant was fabricated for benchtop analysis. Given the prototype nature of this project, funds were minimal and a pilot study was conducted to address feasibility of the expansion anchor approach. To minimize costs and ease machining fabrication the implants were machined from 260 brass. In addition, it was necessary to machine Kurosaka screws from 260 brass. This was done to eliminate material variability in the benchtop testing comparisons, Figure 8.4. To exactly duplicate the dimensions of the Kurosaka screw an optical comparator was used. This device projects an enlarged shadow of the object on a flat surface and measurements can be made based upon a dimensioned template attached to the comparator. In this manner precise thread pitch and thread shape were duplicated. Furthermore, only one initial anchor design was tested. From the results of the benchtop analysis, additional models were sequentially designed and fabricated to address perceived deficiencies in each previous design.

Given that the anchor did not previously exist it was not only necessary to design and fabricate the prototype, it was also necessary to develop the tooling to insert and pull out the anchor and screw designs, Figure 8.5. It was desirable to utilize the mechanical loading machine to standardize testing protocols and eliminate human variability in comparison to using a manual expansion technique. For insertion of the anchors, a standard platen provided the base plate in the testing machine. A steel anvil was fabricated to hold the anchor on the platen and a steel force was fabricated to expand the anchor. The force

Figure 8.4: Brass anchors and screws were compared to eliminate material variability, reduce machining expenses and evaluate design concept.

was designed with three flats to fit into a three-jaw drill chuck inserted into the crosshead (ram) of the loading machine, Figure 8.6. The outer contour of the force was designed to identically mimic the inner contour of the anchor with respect to shape but was oversized to result in expansion of the anchor upon insertion. A testing protocol using the testing machine was developed to run the

force into the anchor and expand it. A loading limit was used to shut the machine off when the force, anchor and anvil were all in contact and the steel force contacting the steel anvil through the anchor resulted in a sharp increase in the load (metal on metal). This was sufficient to permanently set the anchor in the expanded state. Results of these tests were used to verify that a testing machine protocol could be used to successfully expand the anchors while eliminating human variability associated with a manual placement technique. The anchors were expanded with the force at a rate of 0.2 mm/sec. A typical expansion curve is provided, Figure 8.7.

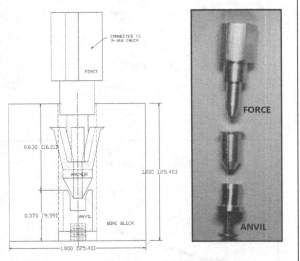

Figure 8.5: Fixtures devised to expand anchor in test block using loading machine for reproducibility.

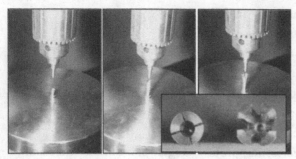

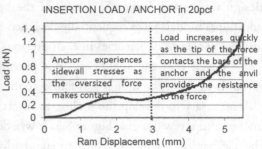

INSERTION LOAD / ANCHOR in 20pcf

Anchor experiences sidewall stresses as the oversized force makes contact

Load increases quickly as the tip of the force contacts the base of the anchor and the anvil provides the resistance to the force

Figure 8.6: Expansion system utilized a base platen that held the anvil. The anchor rested on the anvil and was expanded using the force held in the chuck.

Figure 8.7: Expanded state of anchors with a typical insertion load-displacement curve.

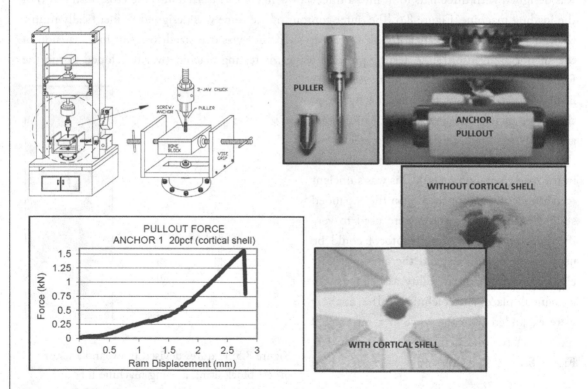

Figure 8.8: Vise fixture was used to hold the bone block while the chuck held the puller that attached to the anchor. Typical load-displacement curve collected from bone blocks with and without cortical shells.

Once a procedure and protocol were developed to reproducibly open the anchors and insert them in a synthetic bone block, a protocol and fixturing were needed to reproducibly pull the anchor from the block and determine holding load of the brass anchors in comparison to the brass screws. Given that the brass anchors were softer than the biomedical metals (stainless steel, titanium), it was not appropriate to use an animal or cadaveric bone block which would be much too hard for this scenario. Instead, synthetic bone blocks (Sawbones) were purchased. Initially a softer cancellous model was utilized that mimicked osteoporotic bone. The softer bone block enabled the testing of the holding load of the systems, whereas cadaveric bone block would have easily resulted in the anchors being bent and damaged upon placement.

In addition to determining an appropriate test block, it was necessary to design a fixturing system to enable the anchors/screws to be pulled out of the blocks while recording the load and displacement. Again, the testing machine was utilized, Figure 8.8. A vise was developed to attach to the base of the loading machine and hold the bone block on two parallel faces. A "puller" was fabricated from 316L steel and threaded to accommodate an internal hole that was drilled and tapped through the end of the anchors and screws. The puller screwed into the anchor while the opposite end had three flats machined to fit into the ram of the loading machine via the three-jaw chuck. Again, the testing machine was used to record load displacement as the screws/anchors were pulled from the block. The systems were pulled out at a rate of 0.2 mm/sec. Typical pullout loads are plotted in Figure 8.8.

Following testing of the first design, it was proposed that a longer anchor with more barbs could be advantageous as it would bite into the bone wall and possibly provide better holding strength. A longer implant was designed which also required the design and fabrication of a new force and anvil to expand the anchor, Figure 8.9. Upon inspection it was immediately evident that the longer anchor in its expanded state could be easily collapsed with a slight load on the open end. While this may have been an issue only with the brass material, given the obvious weakness of the design, it was thought beneficial to design an interlocking wedge system to maintain the anchor in its expanded state and keep it from collapsing while in

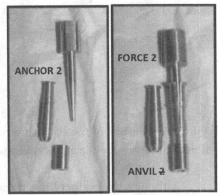

Figure 8.9: Force and anvil developed to test the second, longer anchor design.

use (anchor 3). Results from pullout testing of the longer designs showed little improvement in comparison to the shorter design and with the added concern of the lack of permanent set in the longer design, a shorter design with additional barbs was developed (anchor 4). However, given the long-term use of this system in the loaded environment, the wedge was thought to be a necessary addition. Therefore, the fourth anchor design was developed to incorporate a shorter body, additional barbs and the interlocking wedge, Figure 8.10.

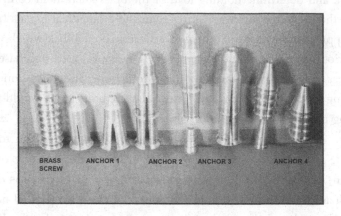

Figure 8.10.: Series of anchor designs that were tested and the brass screw fabricated for comparison.

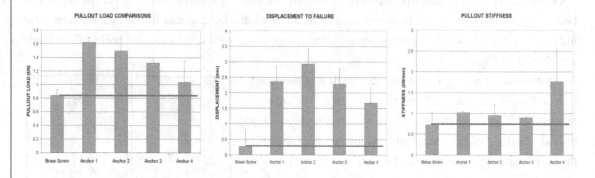

Figure 8.11: Results of mechanical testing. Comparisons of pullout loads, failure displacements and pullout stiffnesses for the anchors and screw.

Although the budget limited the number of anchors for each design that could be fabricated, preliminary mechanical studies were completed and analyses were conducted. From the load-displacement data, the failure load (holding force), failure displacement and linear stiffness were determined. The variability was high but not unexpected given that the sample numbers per design ranged from 8 (anchor 1) to 3 (anchor 4). While rigorous statistical analyses were not undertaken given the small sample sizes, bar charts were developed to qualitatively compare the results, Figure 8.11. Pullout loads for the anchors were at least as strong as the average pullout loads for the screws. Failure displacements were feasible given the manner in which the two systems failed. And, the linear stiffness of the anchors was at least equivalent to the average linear stiffness of the screw.

From a qualitative assessment it was determined that the smaller design performed relatively well in pullout studies. Once it was determined that the expansion approach was relatively sound attention was turned to the issue of size.

To be useful in its intended purpose, the goal of the project was not necessarily to design a new way to anchor a graft, but was to develop a way to increase visibility during surgical placement. The solution chosen here necessitated the design of an expansion anchor. Re-evaluating the previous designs, what was developed was anticipated to be a suitable anchor. At its widest point the implant was 5 mm across, Figure 8.12.

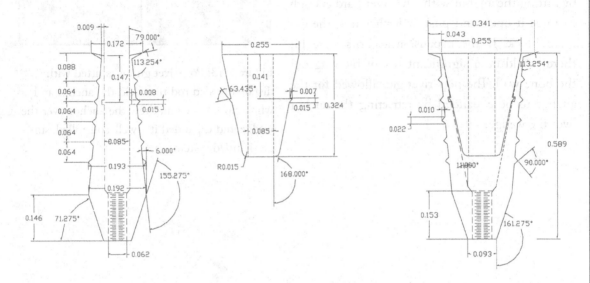

Figure 8.12: Anchor design 5 was arrived at following evaluation of the mechanical performance of anchors designs 1–4. Anchor design 5 consisted of the shorter implant with multiple barbs and a wedge to maintain the anchor in the expanded state.

In addition to the initial anchor design concept, it was also necessary to address placement concerns not only with respect to visibility but also with respect to the development of surgical placement tooling. Given that these procedures are often performed arthroscopically with tension applied to the graft, ideally the placement should occur using a single-handed approach. Therefore, in this benchtop model it was desirable to verify that single-handed placement was feasible. While the solution to enable single-handed placement in the benchtop model would not be the same tooling needed in the operating room, it would verify that these systems could be successfully placed with one hand and adequately demonstrate that a solution for developing surgical tooling exists.

Again, as it is always best to search for a simple solution, one presented itself—a pop-rivet gun and tooling were developed to adapt the anchor and wedge to the gun. A threaded rod was used to accommodate the internal thread of the anchors (the same approach used for attachment to the puller) and a through hole was drilled in the wedge. The threaded rod fit onto the end of the pop-rivet gun, Figure 8.13. In addition to solving the problem of single-handed insertion the pop-rivet gun also enabled constrained expansion of the anchor. That is, cement anchors are expanded by hitting the anchor with a hammer hard enough to force it open and push it further into the cement block. If the expansion anchors were hit there would be a significant risk of blowing out the bone wall. The pop-rivet gun allowed for the anchor to be expanded by retracting the anchor over the wedge.

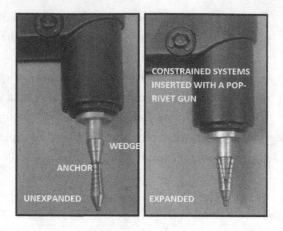

Figure 8.13: Pop-rivet gun was fitted with a distal-threaded rod that held the anchor and when closed, the gun drew the anchor over the wedge and expanded it a well-controlled, single-handed system.

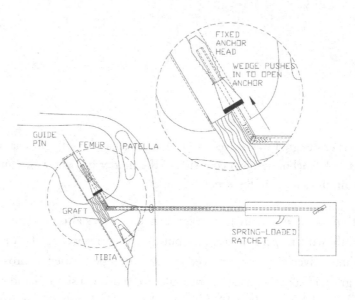

Figure 8.14: Schematic suggesting that suitable surgical tooling could be developed to insert the anchor in a surgical setting.

While the pop-rivet gun would work well for benchtop type analyses a more appropriate option would be needed to address the actual surgical placement of the device. Here a simple spring-loaded ratchet was proposed that could use a chain link mechanism to accommodate the angle and maintain the rigidity needed to push the wedge into the anchor. If this device is ultimately to be successful, surgical tooling would need to be developed to place the device. One idea was proposed here to demonstrate the feasibility of developing the necessary surgical tooling, Figure 8.14. However, given that this was a pilot study, the surgical placement tooling was not further addressed.

An important part of biomechanics is to properly interpret the results of the study and to never extrapolate conclusions beyond what the results directly support. For instance, in this study the displacement for the anchors in comparison to the screw was quite different qualitatively and quantitatively. The anchor had a tendency to "plow" through the bone and continue to displace significantly until the point of failure. As the anchor was pulled out of the bone block, the tension of the system actually continued to force the anchor ends further apart. As is typical of screw fixation, the load was held by the threads and once the hold was broken as the screw displaced and the initial bite was compromised, the displacement at failure was much lower. Other than to comment on the fact that the design readily accounts for the difference, not much else can be said regarding if one failure is better or worse than the other. For instance, where it may appear that the anchor, by continuing to increase in holding strength up to the point of failure is beneficial, the failure is more catastrophic as evidenced by the failure pattern in the bone block (Figure 8.8) which may lead to major concerns if revision surgeries are required for any reason. However, in the living environment, the goal would be for the bone graft tissue to incorporate into the surrounding bone. If incorporation occurred, the displacement may be quite different and given the nature of the synthetic bone block, this study was not designed to address this issue. To address these issues, an animal trial would need to be conducted in which integration had the potential to occur. Pull-out hold could then be assessed after a reasonable time. Furthermore, to conduct an animal trial the brass implants would be unacceptable as the soft nature of this metal would result in bending and damage to the implants as a result of the brass contacting the harder, cortical bone. To conduct an animal study it would be necessary to fabricate the devices from an appropriate metal such as titanium or one of its alloys (Ti6Al4V). As it stands this study was just a first pass at evaluating the possibility of using an expansion device to hold an ACL graft reconstruction. By making both the systems from brass and utilizing identical bone blocks, a comparison based solely upon design was possible. Given that the holding load of the anchor in all designs was at least as strong as the screw, it would suggest that the anchors may perform adequately and additional testing is certainly warranted.

Feasibility tests or pilot study tests are conducted on a smaller scale and used to assess the concept of the project prior to undertaking a full study. In biomechanics, feasibility tests may be used, as demonstrated here, to determine if an idea is worth pursuing on a large scale, or to determine errors in design and performance prior to undertaking a full study. In this case, the full study would not need to only incorporate additional benchtop testing but would be extended, for example to compare screws and anchors from a biologically relevant material such as stainless steel or even resorbable polymers. Additional testing could include sufficient sample sizes to allow for significance with pullout studies and could add fatigue testing. Only after considerable benchtop testing is completed would a design be finalized. Animal trials are expensive and a necessary component of implant development. But they should not be undertaken until a design shows real promise and the use of animals (and their sacrifice) is justified.

In addition to the above issues, additional mechanical testing is also warranted to address mechanical performance of the brass systems. All the first analysis accomplished was to evaluate the holding load of the anchors as a function of their design in a pullout scenario. However, in the body the anchors would experience a multitude of repetitive loads at different magnitudes as a function of knee motion and real-world activities. Therefore, it would be critical to evaluate the performance of these systems in a repetitive loading, or fatigue study. Here the load on the anchors would be cycled between a range of loads and their performance would be compared to that of the screw. An appropriate fatigue waveform, given the results of the pullout tests, may be to cycle the anchors under tension with a waveform that varies between 25% and 75% of the average holding load for a given design. This would be critical to address the ability of the wedge to stay seated in the anchor. As the anchor is repetitively tensioned the effect would also be to separate the ends of the anchor, which if far enough, would allow the wedge to loosen. Given this concern it would be advantageous to utilize the existing internal thread of the anchor and the clearance hole drilled through the wedge (both placed for use with pop-rivet insertion) to run a set or locking screw in the system after placing it in the body to keep the anchor assembly moving as a single unit. It would be wise to undertake these studies in the brass model and, based upon these results, determine if the concept of an expansion anchor is appropriate. If not, the project may be aborted or a new design concept could be developed. While this study described is not complete, as a pilot study it was successful in indicating that there is a potential for these anchor systems in this environment and the results were achieved by the careful development of a study and study protocols to accurately and reproducibly place and pull out the screws and anchors. Using these models and eliminating potentially error-inducing variability, a competent mechanical analysis was completed yielding reliable results.

An important limitation of this work is to realize that in a living system it would be anticipated that with time the bone plug would integrate within the tunnel and provide additional anchorage for the ACL graft replacement. Benchtop tests in synthetic bone block are insufficient to address this issue and animal modeling would be necessary to study this important aspect.

Given that these tests often involve collaboration, it is important to remember that while the results may or may not be what the collaborator is hoping to find, that is really not your concern. Your goal is to develop a sound study such that the results are irrefutable, and a conclusion, in light of the data, may be reached. Law suits arise when promises of clinical outcomes are not achieved. There are no shortcuts when designing implants. Thoroughly and meticulously document all the work.

8.1.5   KEY STUDY CONCEPTS

Design projects are generally collaborative projects. Here a surgeon initiated inquiry into the development of a new device for ACL reconstruction. The first key concept to address is understanding what exactly you need to do, which may or may not be what your collaborator is asking you to do. After considerable conversations with the surgeon, it was actually realized that what was needed was not a device based upon inferior mechanical performance, but one that increased visibility and ease of placement to reduce placement complications. While there may have been other ways to address the increased visibility, we did opt to develop a new device that would insert in an unexpanded state and expand to anchor the graft. Additional concepts related to this study are provided below.

- Your job is to develop a study to produce consistent, accurate, and reproducible results. Your results should determine your conclusion, not vice versa.

- As the biomechanical engineer, your job is to develop appropriate testing protocols and tooling/fixturing to address the question the study is designed to answer.

- Whenever possible eliminate human variability by using testing machines and recording real-time data. Any measurable procedure is superior to one that is not measureable and the more relevant qualitative and quantitative data collected in a pilot study the more productive the study. Note the word "relevant"; not all data collected is relevant and it is important to distinguish between the relevant and irrelevant data.

- Do not extend your conclusions beyond what the study design can support.

- Pilot studies provide an excellent opportunity to develop and verify the adequacy of testing protocols and tooling prior to undertaking a large study. Taking the time to develop a sound mechanical pilot study greatly increases the likelihood that the large scale study will yield reliable findings that will enable identification of design acceptability or necessary design revisions. Mechanical testing should be viewed as a powerful tool to determine reliable results and should not negatively affect what you are trying to measure. A poorly designed study can negatively affect study results and erroneously lead to the conclusion that a sound idea is not worth further pursuing.

- Pilot benchtop study is not sufficient to warrant animal testing, just proof of concept prior to undertaking a larger, more thorough benchtop study. The more thorough study would address issues related to additional testing modes (e.g., fatigue) and enable significance to be established (sufficient sample sizes). Here the pilot study was used to provide proof of concept of an expansion anchor approach and attempted to systematically arrive at a potential anchor design that may be worth further investigation.

- Design prototype studies work to find an acceptable solution to a given problem. The design approach taken here illustrates one solution to the problem. Other approaches could also be developed and this work here in no way implies that it is the best or only solution.

8.2   EXAMPLE 2: CADAVERIC COMPARISON OF ALLOGRAFT FIXATION TECHNIQUES

William M. Parrish, M.D.

Marnie M. Saunders, Ph.D.

8.2.1   THE PROBLEM

When bone is removed as a result of disease (e.g., tumor) or cannot be pieced together after an injury, a skeletal defect is created. In treating skeletal defects in long bones where a sufficient amount of bone is lost, allografts are a common source of replacement tissue. To anchor these grafts in place, plates and nails are used but there does not seem to be a consensus on the most appropriate fixation technique. Moreover, given the variety of complex loading demands placed upon the body there may not be one technique that is optimal for all loading cases. However, regardless of approach the technique must securely hold the allograft in place. While encouraging new bone to integrate with the existing bone, initially the entire load must be borne by the fixation system chosen. That is, initially the post-operative integrity of the allograft-host junction and of the allograft as a whole depends on the stability provided by the device for fixation. The purpose of this study was to mechanically evaluate two commonly used allograft fixation techniques with respect to initial stability at the allograft-host bone junction.

8.2.2   GOAL

To assess the strength of two commonly used femoral graft fixation techniques with respect to initial stability and loading mode.

8.2.3   SOLUTION

For this problem, twenty-one pairs of cadaveric femurs were used. To simulate the graft, the bones underwent a complete mid-diaphyseal transection and the newly created defect simulation was fixed with one of two common fixation techniques—either two bone plates or one intramedullary (IM) nail (running through the medullar bone canal) augmented with one bone plate. Bones were tested to failure in either torsion or three-point bending.

As previously noted, orthopaedic biomechanics projects are commonly done in conjunction with engineers and surgeons. It is generally the surgeon that will bring the idea to the table and it is often inspired by issues that arise in their practice and/or in the operating room. In this case, this study was initiated by an orthopaedic oncologist who used these two techniques (the double plate or IM nail + plate) in his practice and wanted to determine if one technique in comparison to the other provided greater inherent stability as a function of loading mode.

In the body, bones are subjected to a variety of different loading modes including tension, compression, bending, shear and torsion. At any given moment a bone may be subjected to a combination of loads. From a mechanical standpoint this may overly complicate bone loading studies and it often may be just as productive and insightful to separate out the dominant modes of loading and test them individually. With respect to the transected femur model held in place by plates or an IM nail and plate, the system may be particularly prone to failure as a result of bending and torsion. To assess the inherent stability of the two fixation systems under these critical loading conditions, bones were separated into groups and tested with respect to fixation method and loading mode.

8.2.4 METHOD

Table 8.1: Six groups differing by type of fixation and testing protocol

Group	Mode	Fixation	Sample Size
I	Torsion	2 Plate	7
II	Torsion	IM Nail + Plate	7
III	Torsion	Control	7
IV	Bending	2 Plate	7
V	Bending	IM Nail + Plate	7
VI	Bending	Control	6

To verify the appropriateness of the models, it was necessary to verify that they were not in any way diseased and could be used for the study. Twenty-one pair of cadaveric femurs with no visual or radiological evidence of disease were obtained from 15 males and 6 females. Forty-one femurs were tested with one lost to an instrumentation error. Specimens ranged in age from 16 to 83 years of age (mean 51.5) and were donated from the Musculoskeletal Transplant Foundation (Edison, NJ). Cadaver usage was reviewed and approved by the University's Institutional Review Board. On the day of testing, specimens were stripped of all soft tissue and assigned randomly to one of six groups with the groups differing by type of fixation and testing protocol, Table 8.1. Groups I–III were tested in torsion; Groups IV–VI were tested in three-point bending. Prior to fixation, a complete mid-diaphyseal transection was performed in each treated femur and secured with one of two fixation techniques. Groups I (n=7) and IV (n=7) consisted of a double plate construct utilizing both a 6- and 7-hole 4.5 mm Low Contact Dynamic Compression Plate (LC-DCP) (Synthes, Waldenburg, Switzerland) placed orthogonally on the anterior and lateral cortices, respectively and secured with 4.5 mm self-tapping bicortical screws, Figure 8.15. Groups II (n=7) and V (n=7) consisted of a statically locked 9 mm diameter IM nail (Synthes, Waldenburg, Switzerland) and a 7 hole LC-DCP placed laterally and secured with unicortical 4.5 mm self-tapping screws to provide compression and rotational control, Figure 8.15. Groups III (n=7) and VI (n=6) were non-instru-mented, intact controls. In all cases, an attempt was made to test the specimens within 18 hours of fixation. In the event that this was not possible, specimens were refrozen in saline-soaked gauze until use. Care was taken to ensure that no specimen underwent more than two freeze/thaw cycles and specimens were thawed and rehydrated in room temperature saline prior to biomechanical test-ing. Destructive torsion and bend testing were completed on a biaxial closed-loop servo-hydraulic materials testing machine (Interlaken, Eden Prairie, MN) utilizing a 10 kN load cell and a 400 Nm torque cell. Testing and data collection were computer-controlled. Load-displacement and torque-twist data were recorded and fracture location was noted.

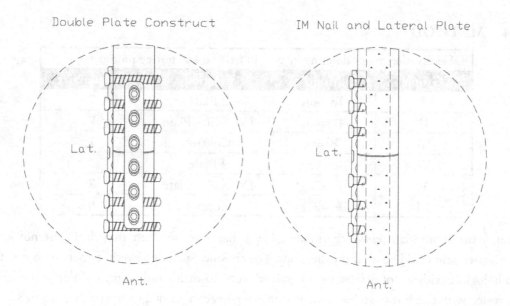

Figure 8.15: Fixation techniques. a) Simulated allograft fixations of Groups I and IV were performed with double plates on the lateral and anterior bone surfaces; b) simulated allograft fixations of Groups II and V were performed with an IM nail and lateral plate.

For torsion testing the specimens were potted on center in square, aluminum potting fixtures using Bondo™ and bone screws. The potting jigs were used to ensure that the specimens would axially align in the square fixtures used for torsional testing. This ensured that the bones were not forced into the testing frame, there was no unwanted pre-load on the specimens prior to or during testing and that a torsional load was applied. The torsion test configuration consisted of two square steel chucks mounted to the load cell and crosshead of the testing machine, Figure 8.16. Prior to testing, the potted specimens were inserted into the fixtures and the load cell was tared. All specimens were externally rotated to failure under rotational control utilizing a linear ramp waveform at a rate of 3 degrees/second with torque-twist recorded. Data analysis for each specimen (Groups I–III) included determination of torsional stiffness taken as the slope of the torque-twist curve in the linear region, failure torque taken as the peak torque, twist to failure taken as the degree at which failure torque occurred and fracture location.

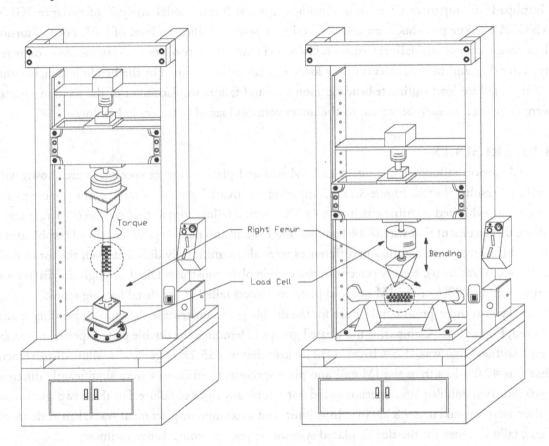

Figure 8.16: Testing schematics for a) torsion and b) three-point bending. The load/torque cell on this machine is a combination cell capable of reading both axial and rotary loads.

For bend testing, specimens were placed on the aluminum, three-point bend fixture affixed to the testing machine with the central contact in line with the diaphyseal transection of the treated groups, Figure 8.16. This simulated the worst-case scenario in which the load was directly applied across the fracture surface. Spacing of the bottom contacts was on center and 17 cm apart to ensure a 2.5 cm space between each of the bottom contacts and the trailing edge of the 7-hole plate. This resulted in a highly concentrated load being placed across the instrumented region. Specimens were loaded to failure under displacement control utilizing a linear ramp waveform at a rate of 1 mm/sec with load-displacement recorded. Data analysis for each specimen (Groups IV–VI) included determination of bending stiffness taken as the slope of the load-displacement curve in the linear region, failure load taken as the peak load, displacement to failure taken as the displacement at which failure load occurred and fracture location.

Statistical analysis was completed with a commercially available software package (Instat, Graphpad). Comparisons were made using a general linear model analysis of variance (GLM ANOVA), Tukey post-hoc comparisons, and an *a priori* significance level of 0.05. For the torsion data, torsional stiffness, failure torque and twist to failure of the two fixation systems were compared to each other and to the intact controls loaded to failure in torsion. For the bending data, bending stiffness, failure load (ultimate bending moment) and failure displacement of the fixation systems were compared to each other and to the intact controls loaded to failure in bending.

8.2.5   RESULTS

Typical torsion failures in the intact and IM nail and plate construct specimens are shown with torsional results plotted, Figure 8.17. Comparing the treated groups in torsion, the double plated specimens exhibited a stiffness that was 43.3% larger, a failure torque that was 36.6% larger and a failure displacement that was 37.0% less than the IM nail and plated specimens. In all combinations these differences were significant; differences were also significantly different from the intact controls ($p < 0.05$, at least). Eighty percent of the double plate constructs failed through the distal most screw hole and 80% of the IM nail and plate constructs failed at the distal locking screw.

Typical three-point bend failure for the double plate construct is shown with bending results plotted, Figure 8.18. Comparing the treated groups in bending, the double plated specimens exhibited a stiffness that was 45.5% larger, a failure load that was 15.2% larger and a failure displacement that was 42.0% less than the IM nail and plate specimens. Stiffnesses were significantly different ($p < 0.05$). Two double plated specimens did not exhibit any signs of failure. For these two specimens, failure load was taken as 9.8 kN (machine limit) and maximum displacement was taken at this load. Thus, failure values for the double plated systems represent a conservative estimate.

Specimens with double plate fixation exhibited larger torsion and bending stiffnesses as compared to the specimens with IM nail fixation. Double plate constructs showed a higher failure load in torque and smaller failure twist as compared to IM nail constructs. Similarly, double plate constructs exhibited a higher failure load in bending and smaller failure displacement as compared to IM nail constructs. These data suggest that allograft fixation with double plating increases the stiffness of the construct and thus offers superior initial stability of the allograft-host bone junction in comparison to IM nail with plate supplementation. The immediate goal of internal fixation in allograft fixation is to provide stability at the host-bone junction to facilitate healing and decrease the risk of subsequent fractures. Study findings support the use of double plate constructs in helping the surgeon achieve better initial stability at the host-bone junction and decreases the risk of allograft fracture over the IM nail and single plate fixation approach.

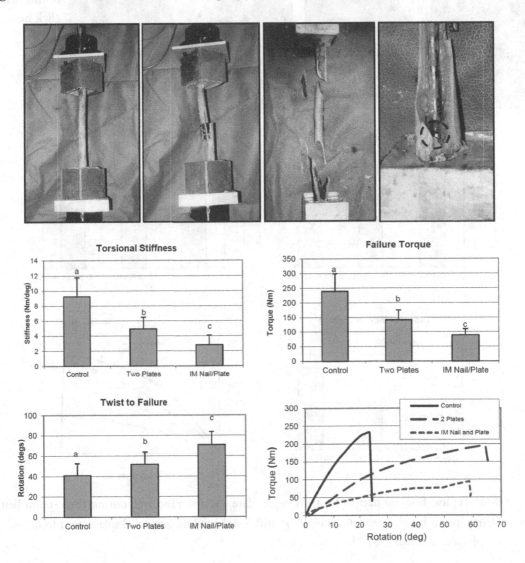

Figure 8.17: Typical fracture patterns for the intact (top) and IM nail and plate construct (bottom) incurred during torsional testing. Torsional results. a) torsional stiffness; b) failure torque; c) rotation to failure. All results are plotted as mean ± SEM.

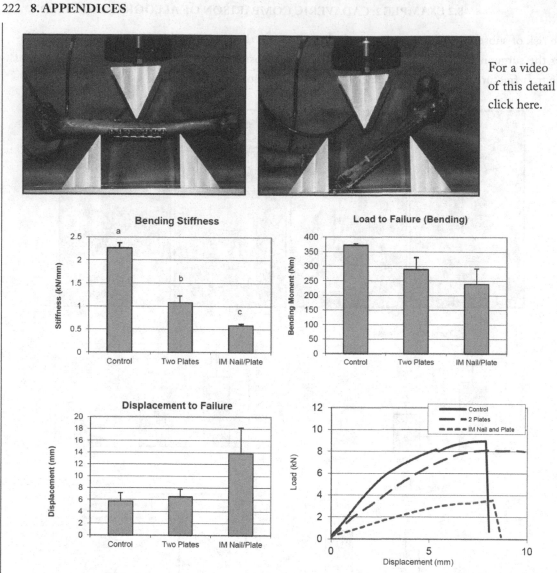

For a video of this detail click here.

Figure 8.18: Typical fracture pattern for the double plate construct incurred during three-point bend testing. Three-point bending results. a) bending stiffness; b) failure load; c) displacement to failure. All results are plotted as mean ± SEM.

8.2.6 STUDY LIMITATIONS

In this study the initial stability provided by double plate fixation and IM nail fixation in a cadaveric femoral model was examined. That is, two standard allograft fixation techniques under two loading modes (torsion and three-point bending) were mechanically evaluated. The method of testing used separates some of the mechanisms in physiological loading by isolating the loading parameters and thereby differentiating the loading conditions experienced by the femur. A significant advantage of the testing method reported was measurement of stiffnesses giving a measure of the overall composite rigidity of the two fixation techniques and the use of a transected cadaveric model simulated allograft fixation immediately post-implantation (no healing period). Therefore this study made no attempt to address postoperative union between the allograft and host bone and the results obtained should not be used to predict union. However rigidity is an important indicator of interface micromotion and thus is an acceptable predictor of post-operative stability.

Factors that should be considered in the evaluation of union include soft tissue damage caused by the fixation technique and the quality and geometry of the host bone. Plate fixation causes significant trauma to the host bone when periosteum and soft tissues are stripped for placement of the plate and bone properties differ with patients. For instance, with increased bone porosity higher failure rate with the use of dynamic compression plate can be observed. This study did not take into account this concern and evaluated the ideal situation with the emphasis toward determining the construct that would provide better stability immediately post-surgical implantation. Moreover fatigue testing may provide important information by simulating dynamic conditions. The goal of the current study was to biomechanically compare two commonly used allograft fixation systems with respect to initial graft stability. It was determined that double plate fixation provided a significant increase in torsional strength and stiffness relative to IM nail fixation with plate supplementation and a significant reduction in torsional twist. Furthermore, the double plate fixation provided a significant increase in bending stiffness relative to IM nail fixation. The low stiffness of the IM nail construct resulted in a lower rotational stability of the femoral shaft as compared to double plate fixation. Taken together these findings suggest that double plating is the superior post-operative fixation system and should clinically facilitate healing while minimizing the risk of nonunion. For the patient, this translates into an expected increase in the longevity of the fixation as well as an expected increase in bone quality and stock around the construct.

8.2.7 KEY STUDY CONCEPTS

As general practice, in biomechanics fixation comparisons ten to twelve specimens per group are usually sufficient to detect a true significance if one exists. A Power analysis may be conducted with the population parameters of the smaller sample size to approximate the number of specimens needed to obtain significance. Finally, if sample sizes sufficiently larger than ten to twelve are needed to obtain significance, one should question the relevance of the study. For example, in the previous study seven samples per group was sufficient to determine significance. If a Power analysis determined forty samples per group were needed, the value of the study should be called into question and reevaluated. That is, if forty samples are needed per group, is the factor the study is designed to evaluate really that relevant?

As noted in the study, two plated specimens did not fail under the bending load. Here failure load was taken as the end of the load cell readout. The test still yielded reliable data and it can be reliably stated that the failure load was "at least as high as" the load cell readout at the time of machine shutdown. In this regard, the results for these two tests are conservative estimates of failure and have the effect of underestimating the mean failure results for the double plated constructs.

It is important to remember that decisions to address fracture fixation are not only based upon mechanical performance. Biological factors should also be considered. For example, if the plates sacrifice the periosteum and affect the long-term healing potential (incorporation of the graft with the native tissue), these factors need to be considered when determining the fixation technique. Furthermore, these decisions may need to be made on a case-by-case basis noting that with age, the regenerative potential of bone decreases. It is always the responsibility of the surgeons to make these decisions. Your job as the engineer is to provide the surgeon with the reliable data with which to base their decision. Furthermore, realize that the data is interpreted and the conclusions may not be black and white. For example, while this study showed that in three-point bending and torsion the two-plate fixation approach was mechanically superior to the IM nail and plate fixation approach, it does not mean that the IM nail and plate fixation approach should never be used. There may be additional factors that contribute to the surgeon's decision and there may be other loading scenarios where this fixation approach is acceptable. In keeping with the idea to not extrapolate your study conclusions beyond what the results support, don't undertake a study with the idea that one approach is "good" and the other is "bad." Undertake the study with the notion of conducting reliable biomechanics work to quantify under the exact confines of the constraints of your study (bone model, fixation systems, loading modes) the performance of each system. The surgeon will interpret the findings in a clinically relevant context. It is the surgeon who will ultimately examine the data and determine the usefulness of a clinical system.

It is important to report any experimental limitations that arise. For example, in the current study a bone was broken during the fixation process. The loss of this bone does not negate the quality of the study or the usefulness of the findings. Any time testing issues arise, you want to make sure to report them so that your work may be correctly interpreted by others in light of any testing concerns. For example, the fact that two double-plated constructs did not break indicates that the loading machine limits were exceeded in this study. Given that the IM nail constructs all broke in the testing ranges of the machine suggest that valid comparisons may still be made. However, double-plate failure values are actually larger than what was reported in this study. If similar fixation systems are reported by other research groups with testing machines with larger load capabilities, the results in this study will not match the work of the other groups. It is important to note these occurrences so the data may be correctly interpreted. Furthermore, if the goal of this study had been to determine the ultimate failure loads associated with double plate fixation, the results here would not adequately address that goal. However, as is often the case with biomechanical fixation studies, the goal is to compare two fixation systems to determine which provides superior stability. Given the relative comparisons inherent in this type of research, the loading machine limitations do not affect the acceptability of the results in this study. If both fixation systems had several specimens that did not fail within the loading limits of the machine, it would not be possible to make conclusions based upon this work.

8.3 EXAMPLE 3: BONE REMOVAL LOCATION EFFECT IN AUTOGRAFTING: ASSESSING FRACTURE RISK

David C. Goodspeed, M.D.

Marnie M. Saunders, Ph.D.

8.3.1 THE PROBLEM

Autografting of bone requires that bone be harvested from a secondary site in the body. These donor sites are generally selected for the quality and volume of bone stock as well as ease of harvesting. The distal femur and proximal tibia provide alternative sources for bone graft harvest. They have the advantages of abundant graft material, minimal post-operative pain and are often in the standard operative field. Although harvesting from this site has been conducted in numerous patients, torsional fracture of the bone has been observed, Figure 8.19. This has prompted the current study to quantify the extent to which the donor site location influences torsional bone strength and whether or not this procedure puts the patient at increased risk of fracture. No biomechanical studies in this area have been previously conducted.

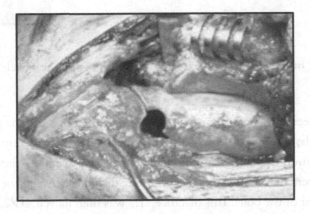

Figure 8.19: Clinical spiral fracture at autograft harvest.

8.3.2   GOAL

To quantify the torsional performance of femurs post graft harvest to determine the appropriate harvest location and if mechanical performance is compromised at that location.

8.3.3   METHOD

Thirty fresh frozen, human femurs were used for this study with IRB approval. All specimens were obtained from the Musculoskeletal Transplant Foundation with the study population comprising femurs from adult men and women donors between the ages of 18 and 55. Twelve hours prior to testing, specimens were removed from the freezer (-20°C), submersed in saline and allowed to thaw at room temperature. Following removal of all soft tissue, specimens were DEXA scanned at Ward's triangle and omitted from the study if DEXA levels indicated osteoporosis.

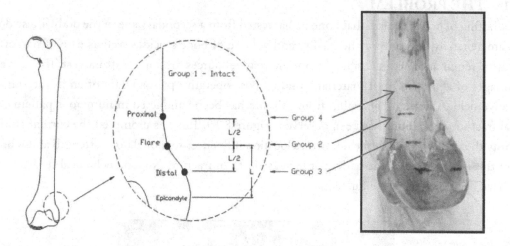

Figure 8.20: Definition of hole locations for treated groups shown in the sketch and marked on the specimen.

Twenty-four specimens were then randomly assigned to one of four groups. Group I consisted of intact controls; Groups II–IV consisted of specimens with 13 mm holes in the anterior, distal portion of the femur. The hole locations were defined as "proximal," "flare," and "distal," Figure 8.20. Hole locations were determined by measuring the distance between the distal epicondyle and inflection ("flare") point of the bone. This distance, "L," was then halved, "L/2." The "proximal" point was designated as L/2 units proximal to the inflection point, the "flare" point was designated as the inflection point and the "distal" point was designated as L/2 units distal to the inflection point. This resulted in a distance between the farthest holes ("proximal" and "distal") of L units apart. Hole placement was performed using guide wires and a 13 mm trephine, Figure 8.21. Trephines were marked to ensure that equal volumes of bone were removed from the harvest site and correlated with the amount of bone the surgeon removes from the patient during the surgical autograft harvest. In addition to the randomized study, three matched pairs of femurs (of comparable age and DEXA properties) were also tested using the same locations and a 10 mm trephine. The right legs were randomly assigned to intact controls and the lefts were assigned to holes at one of the three locations. These data served as a direct indication of the relative loss in strength of the femurs as a function of hole size and were designed to minimize the number of cadavers utilized in the first phase of the study by not requiring matched pair testing throughout.

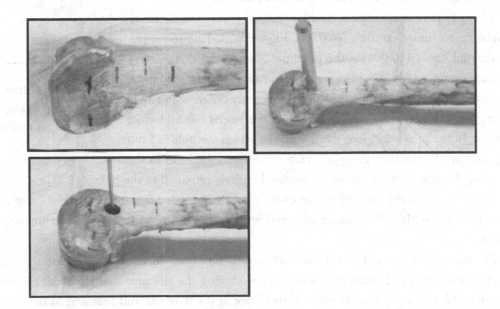

Figure 8.21: Location of the harvest sites. Using the flare and epicondyle of the distal femur as reference points, the proximal, flare and distal sites were marked. A Steinmann pin was then inserted as a quide wire and the trephine was then used to place the harvest hole and the bone stock removed. In this way, reproducible holes were placed in the treated groups.

For torsion testing, specimens were potted on center in custom-designed, square, aluminum chucks using commercially available dental cement (Duz All, Harry Bosworth, Co), Figure 8.22. This was done to ensure that the specimens would axially align in the square fixtures used for torsion testing. Specimens were potted at or below the level of the epicondyle and anchored in the cement with two distal bone screws. Specimens were torqued to failure in a closed-loop servo-hydraulic testing machine (Interlaken). Prior to testing, the potted specimens were inserted into the fixtures and the load cell was tared. All specimens were externally rotated to failure using rotational control and a linear ramp waveform at a rate of 3 degrees/second with torque-twist recorded.

Data collection and analysis for each specimen (and each treatment) included torsional stiffness data taken as the slope of the torque-twist curve in the linear region, failure torque data taken as the peak torsional load, failure twist data taken as the angle at which the peak torque occurred, and fracture location. Data was grouped and analyzed using standard GLM ANOVAs with SNK post-hoc comparisons and an *a priori* significance level of 0.05 for all comparisons.

Phase I: Torsional stiffness of the proximal group was decreasd 37% relative to intact controls (p<0.05), Figure 8.23. Failure torque was decreased 46% (p<0.001) in the proximal group and 19% (p<0.05) in the flare group relative to intact controls but was not affected in the distal group. Rotation to failure was not significantly different in any of the four groups. Fracture location in the intact specimens was mid-shaft in all cases. In the proximal group the fracture traversed the hole in 5 out of 6 specimens. In the flare group all fractures went to the hole. In the distal group 5 out of 6 fractures were contained entirely proximal to the hole.

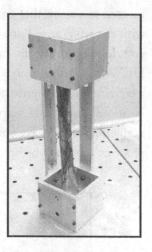

Figure 8.22: Potting fixtures used to align specimens for torsional testing.

Phase II: The torsional stiffness in each of the three treated femurs was within 7% of intact; failure torque was within 5% of intact and rotation to failure was not affected. All fractures occurred mid-shaft.

Placement of a lateral 13 mm bone graft harvest hole does not compromise torsional strength when placed at the distal location defined above. Moving the site proximally reduces mechanical performance (failure torque). This effect is moderate at the flare site and dramatic at the proximal site, correlating with our clinical observations. Furthermore, reducing the size of the harvest hole to 10 mm restores the structural integrity to that of intact controls. Based on the above findings, we continue to use the distal femur as an alternative harvest site by placing the harvest hole at or below the flare site. No complications to date with this constraint have been observed.

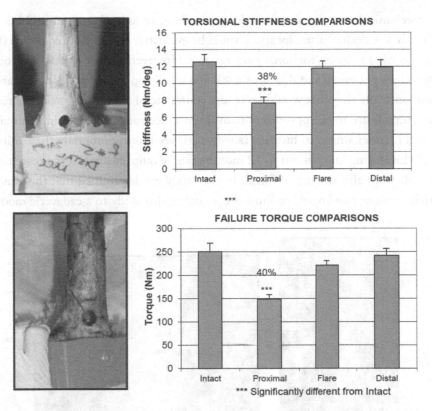

Figure 8.23: Torsional results. Torsional stiffness and failure torque were significantly reduced in the proximal group using the 13 mm trephine. Results are plotted as mean±SEM.

8.3.4 KEY STUDY CONCEPTS

Much of biomechanics work is understanding that theoretically procedures such as this, bone removal, have an effect on bone strength. The testing will determine if the effect is functional. That is, does the procedure affect the bone to the point that it initiates or expedites failure?

Cadaver models are not always the ideal choice for quality assurance type fixation studies. When they are very appropriate is when a healing stage is not incorporated. Given the use of dead, cadaver tissue the tissue will not regenerate. In the clinic, the procedure investigated in this study would ultimately result in healing and the filling in, at least in part, of the bone removed, Figure 8.24. Since the goal of this study was to determine if the harvesting procedure compromised the mechanical integrity of the femur, the cadaver model (no healing potential) provided a worst-case scenario in which to assess the mechanical integrity if new bone formation/hole filling did not occur. This study does nothing to address mechanical integrity in the presence of new bone growth. This is another example of where multiple studies are needed to thoroughly address a problem. To address the mechanical integrity as a function of healing, an animal study would be required. In this type of study the desired distal location would be evaluated at various time points (for example 1,3,6 and 12 months) to assess torsional performance. However, a study of this size could be quite large (10 specimens per group and 4 groups = 40) and ideally only one femur in each animal would be operated upon while the other would serve as a contralateral control. To make the results meaningful, larger animals are more appropriate than for instance rodent models and animals that load their bones in a manner similar to humans (sheep, pigs) are more appropriate than animals that do not (rabbits). Monitoring animals out to 12 months can be quite costly and the researcher needs to determine if this is an effective use of funds. In this study we determined that this was not a good use of research resources and opted to limit the model in this study to a cadaveric model.

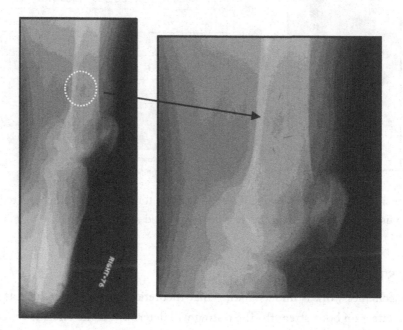

Figure 8.24: MRI results of clinical follow-up 18 months post-operative.

In addition to the study to assess the harvest location, three cadavers were used to assess the effect of using a smaller trephine size. Not all surgical procedures are thoroughly evaluated prior to clinical use. In this case, the harvesting procedure had been used in patients (a minimal number). It was not until a patient fell and suffered the fracture that the surgeon began to question the safety of the procedure. This technique is being used and biomechanical data helps to assess the acceptability of the technique and provide guidelines for location and safety.

8.4   EXAMPLE 4: DIAPHYSEAL FEMUR FRACTURE AFTER PROXIMAL AND DISTAL FIXATION

J. Spence Reid, M.D.
Marnie M. Saunders, Ph.D.

8.4.1   THE PROBLEM

Studies have demonstrated the importance of spanning a cortical defect by at least two bone diameters (BD) in joint arthroplasty. However, it is unclear if this applies to fracture fixation. Theoretically, if proximal and distal fixation were close enough a critical length of exposed bone would be maximally vulnerable. Assuming a stress riser effect exists between proximal and distal femoral fixation, our objective was to determine the amount of uninstrumented bone that results in a significantly weakened region of diaphyseal bone.

8.4.2   GOAL

This study examines the amount of femoral shaft left unprotected that puts the femur at risk for fracture between the fixation devices. In theory, if the proximal and distal fixation devices were far enough apart, the stresses to the unprotected midshaft region would resemble that of normal bone. Conversely, if the proximal and distal fixation devices were close enough together, there should exist a critical length of unprotected shaft region that would be maximally vulnerable to stress riser effects and subsequent fracture. Here we wanted to ascertain whether an effective stress riser exists between proximal and distal fixation sites on the adult femur, whether this effect could be influenced by varying the amount of unprotected bone between fixation sites and whether spanning the unprotected region results in a decreased risk of fracture.

8.4.3   METHOD

Forty-eight sawbone femurs underwent proximal hip screw fixation and distal intramedullary nailing. Nail lengths were varied to overlap or lie tip-to-tip, 1, 2.5 or 4 bone diameters from the side plate, and proximal nail fixation varied to test fixation through proximal, distal or both locking sites. Femurs were tested to failure in three-point bending. Force, displacement, stiffness and fracture patterns were examined.

Neck/shaft femur fractures present a unique problem for the orthopaedic surgeon. Approximately 5% of high-energy femoral shaft fractures will be associated with an ipsilateral femoral neck fracture. Additionally, with the increasingly active geriatric population, orthopaedic surgeons may be treating higher numbers of femur fractures occurring after previous femoral fixation. Proximal and distal fixation by a combination of intramedullary nails, plates and screws leaves the intact unprotected segment of the shaft at risk for fracture due to a putative stress riser effect between the proximal and distal devices. In order to avoid this effect many surgeons resort to long intramedullary nailing that spans the unprotected region. However no studies have been performed that examine the effectiveness of this procedure, nor have any studies examined if the putative stress riser is functionally relevant.

Little is known about the forces upon a femoral shaft after proximal and distal fixation. Previous studies have concentrated on examining the stress riser effect of an intramedullary prosthesis upon the surrounding bone. Courpied et al. demonstrated that in the case of ipsilateral total hip and knee replacements in which the femoral component of the knee prosthesis uses and intramedullary stem there is a greater concentration of stress between the tips of the hip and knee components. However, few studies to date have examined how proximal and distal fixation of femoral fractures affects the intervening segment of femoral shaft. Numerous methods have been developed to avoid fracture in this area by early hardware removal, spanning the area with a long plate or a long intramedullary nail or if possible, maximizing the distance between the proximal and distal hardware. While studies have shown the importance of spanning a periprosthetic fracture by at least the distance of two bone diameters, bone strength is 80% of normal at this distance. However, it is unclear whether this principle applies not only to periprosthetic fractures around a solid intramedullary prosthesis but also to normal bone adjacent to an area that has been augmented by hardware fixation.

8.4.4   THE SOLUTION

Forty-eight large (adult) third generation mechanical composite sawbone femurs (Pacific Research Laboratories, Vashon, WA) were divided into eight groups. Seven groups underwent proximal placement of a 95 mm DHS sliding hip screw through the femoral neck into the head followed by fixation to the proximal femoral shaft with a 135-degree 4-hole DHS side plate (Synthes, Warsaw, IN). Screws were placed through all four holes of the side plate. This was followed by placement of a titanium retrograde intramedullary nail (Distal Femoral Nail System, Synthes, Warsaw, IN). Nail lengths differed for each group such that the proximal end of the nail was 0 ("tip to tip"), 1, 2.5 or 4 bone diameters distal to the end of the DHS side plate. One group had overlapping fixation such that the intramedullary nail overlapped with the side plate by 1 bone diameter, and one group was left uninstrumented ("control"). All retrograde femoral nails were distally locked with two lateral-to-medial locking screws and an end cap. Five of the groups were proximally locked in the femoral diaphysis with one anterior-to-posterior locking screw placed through the locking site closest to the hip. The remaining two groups consisted of nails that were 1 bone diameter from the end of the side plate and were fixed with one anterior-to-posterior locking screw through the locking site closest to the knee or with two anterior-to-posterior locking screws through both locking sites in the nail, Figure 8.25.

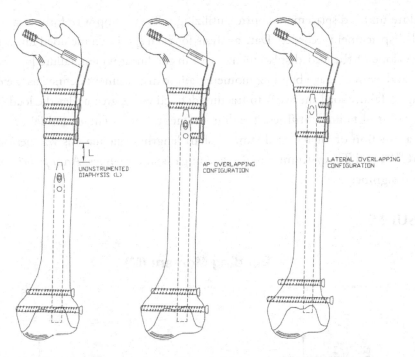

Figure 8.25: Fixation configurations tested in this study.

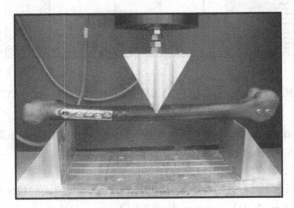

Figure 8.26: Instrumented femur tested to failure in three-point bending.

The femurs were loaded to failure in three-point bending using a biaxial servohydraulic testing machine (EnduraTec, Minnetonka, MN). Each femur was place on an aluminum (T6061) three-point bending fixture with the central contact in line with the mid-diaphyseal region and the anterior femoral cortex under tension and the posterior femoral cortex under compression (Figure 8.26). Spacing of the bottom contacts was on center and 30.5 cm apart. Specimens were loaded to failure under displacement control, utilizing a linear ramp waveform at a rate of 1 mm/sec with load-displacement recorded. Data analysis for each specimen included linear stiffness data defined as the slope of the load-displacement curve in the linear region, failure load defined as the ultimate load and expressed as a bending moment, failure displacement defined as the displacement corresponding to failure load and work to fracture defined as the area under the load-displacement curve (to the point of failure). Stiffness, bending moment, load to failure and failure location were examined as a function of unprotected femoral shaft length. Data analysis was performed using a one-way ANOVA and Tukey-Kramer multiple comparison post-hocs. An *a priori* p value of 0.05 was considered significant.

8.4.5 RESULTS

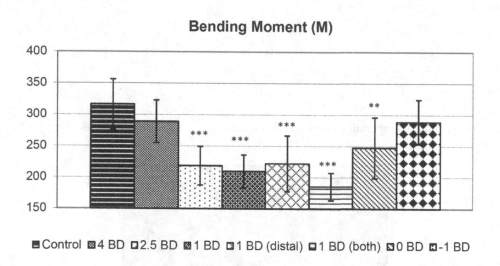

Figure 8.27: Bending moment comparisons of instrumented femurs and control. *** p<0.05 vs control, 4 BD and 1 BD; ** p<0.05 vs. control only.

Moment to failure was significantly decreased with 1–2.5 bone diameters (BD) of diaphysis exposed. Femora with less than 1 BD or more than 4 BD exposed demonstrated loads to failure that were not significantly different than control (Figure 8.27). All fractures occurred as short oblique fractures through the proximal locking screw of the nail, except where IM nail fixation overlapped the proximal plate or was more than 4 BD from the proximal fixation. In these cases fractures occurred transversely at the diaphyseal region closest to the area of highest bending force (Figure 8.28). The addition of instrumentation did not significantly alter the load to failure. Stiffness in all groups was not affected by placement of instrumentation or instrument configuration (Figure 8.29).

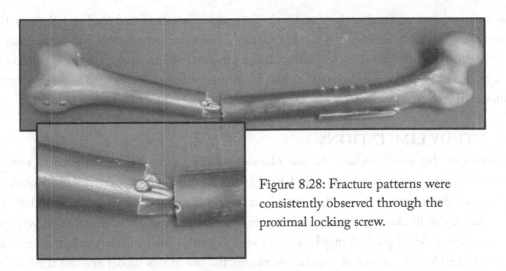

Figure 8.28: Fracture patterns were consistently observed through the proximal locking screw.

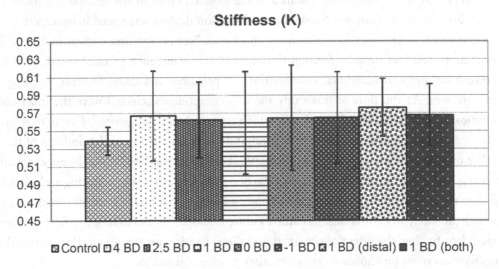

Figure 8.29: Stiffness comparisons of instrumented femurs and control. No statistically significant differences were detected (p<0.05).

The two groups that underwent proximal fixation through both locking sites or the locking site closest to the knee were examined for differences in fracture location and load to failure. As demonstrated, no differences were seen in failure moment at 1 BD with respect to noninstrumented controls. Fractures occurred through the most proximal locking screw regardless of the number of locking screws or the distance of the screw from the tip of the nail. Failure moment was not significantly different from noninstrumented controls by placing fixation more than 4 BD apart or by overlapping fixation by 1 BD. Interestingly, placing fixation "tip to tip" with the proximal end of the nail at the same level as the distal end of the side plate increased load to failure such that this construct was only significantly weaker than noninstrumented controls and not weaker than any instrumented construct. This was also true for proximal nail fixation through the locking site closest to the knee in the 1 BD construct. However, this increased failure load was not seen when proximal nail locking sites were used.

8.4.6   STUDY LIMITATIONS

Weaknesses of this study include the use of sawbone composite femurs designed to reproduce the density of the adult male femur and would simulate mechanical performance immediately post-fixation. Such femurs may not accurately represent findings seen clinically, especially older, more sedentary individuals in which dual fixation may be utilized. In addition, sawbones cannot undergo bone remodeling which may be a factor seen in femurs having undergone previous fixation. While not ideal in comparison to a cadaveric model, the use of this model enabled the large population size, multiple fixation comparisons and the isolated effect of the mechanical fixation to be evaluated. For this study, only one combination of fixation devices was tested in order to specifically address the stress riser effect in the femoral diaphysis. Newer instrumentation including locked periarticular plates, less invasive fixation devices and the resurgence of trochanteric nail fixation also present unique and individual combinations of proximal and distal fixation that may create different stresses. At the time of this study the two components selected were the most common combination seen in the clinical practice in which this study idea originated. Future developments and advancements in instrumentation may precipitate a reexamination of the findings of this work.

The use of three-point bending as the method of failure was based upon biomechanical studies showing that three-point bending most closely reproduces stresses seen during ambulation. This would best reflect insufficiency fractures in bone and may not accurately reflect forces seen during falls, which are likely to be torsional in nature. Future torsional studies examining failure torque may help elucidate further the effect of proximal and distal fixation on the femur. But, as noted above. potting/fixation must be chosen so as not to alter mechanical analysis.

In the review of the literature prior to undertaking this study, only one study attempted to examine the effect of proximal and distal fixation on uninstrumented bone. In a cadaveric study, Harris et al. demonstrated that overlapping fixation serves to increase the load to failure in bone subjected to torsional load when compared to femurs in which the fixation devices were placed more than 0 BD apart. Our results support this finding but further show that the amount of bone between devices has a large influence on failure load. Our data suggest that a distance between fixation devices of at least 4 BD results in failure loads similar to uninstrumented intact bone. Furthermore, a particularly vulnerable area for fracture appears to be through the proximal locking screw of a retrograde nail when the tip of the nail is 2.5 BD or less from the tip of a sliding screw side plate.

8.4.7   KEY STUDY CONCEPTS

Proximal and distal fixation by a combination of intramedullary nails, plates or screws leaves the intact unprotected segment of the shaft at risk for fracture due to a putative stress riser effect between the proximal and distal devices. One solution that avoids the problem is to use a second generation antegrade nail. However, this device can be challenging to place correctly. Additionally the increasing geriatric population is at risk for insufficiency femoral fractures proximal or distal to preexisting fixation devices in the femur. These situations may mandate the use of two separate devices. Few studies have specifically examined the amount of femoral shaft left unprotected that places the femur at most risk for fracture between fixation devices. This study utilized mechanical composite sawbone femurs to minimize variability in bone density, geometry and anatomic differences. Three-point bending was used to simulate normal forces seen in ambulation with tensile forces directed on the anterior cortex and compressive forces directed on the posterior cortex. An added advantage of this testing approach was that it did not require potting/fixation of the bone ends such as is required for torsional and axial testing, which could result in end effects.

The study demonstrated that instrumentation of the femur results in mechanical performance that can be influenced by the configuration of the fixation devices and is weakest when the devices are 1–2.5 BD apart at their proximal and distal ends. This effect can be ameliorated by either changing the distance between devices or by placing the retrograde nail proximal locking screw in the hole closest to the knee. Configurations in which the fixation devices overlapped, lay tip to tip, or lay 1 BD apart with only the distal locking site used were not significantly different with respect to failure load. Also, overlapping fixation was not as strong as when the fixation devices lay 4 BD apart. Thus, previous attempts to "shield" diaphyseal bone by overlapping fixation may not have provided the benefit originally thought. Also we found that in the weaker fixation configurations, the fracture inevitably occurred at the proximal locking screw of the retrograde nail. Thus, proximal fixation may create a stress riser effect not previously noted.

8.5   EXAMPLE 5: CELLULAR BIOMECHANICS: MECHANICAL PLATFORMS FOR MECHANOBIOLOGY

Marnie M. Saunders, Ph.D.

Some may argue that much of what is classically considered biomechanics is, in fact, a dying science. Much of what we classically use biomechanics for, such as determining the mechanical properties of tissues, particularly bone, has been accomplished. In contrast, there are still considerable applications in cardiovascular biomechanics. While this is a matter of opinion, it cannot be denied that biomechanics is evolving and researchers that want to be invested in the field will need to adapt. For instance while there continues to be a place for biomechanics in orthopaedic research, for instance in testing scaffolds, transgenic animals and animal systems post-healing, there are also tremendous opportunities in the field of mechanobiology which acknowledges that mechanics play an important role in the development, function and pathology of bone.

8.5.1   THE NEED

In addition to the classic biomechanics studies conducted on tissues, research into mechanobiology and more specifically mechanotransduction is aimed at understanding how bone cells sense and respond to mechanical loading. Because bone cell response may be measured directly at the cellular level by quantifying cellular activity or indirectly at the tissue level by quantifying bone formation/resorption (the product of cell activity), both *in vitro* (cell) and *in vivo* (animal) mechanotransduction models exist. *In vitro* mechanotransduction platforms simulate the loading the cells experience in the physiologic environment and attempt to elucidate the pathways and mechanisms by which the cells respond. Brown has written an excellent review of these *in vitro* systems and their advantages and disadvantages. Here we will justify the biological premise for fluid flow and substrate deformation.

In the native bone tissue osteocytes are housed in lacunae in the bone matrix and form a networked conduit, the lacuno-canalicular system, with other osteocytes via canalicular channels that connect the lacunae. This unique network enables osteocyte processes to physically connect in a shared pool of interstitial fluid that bathes this complex. The osteocytes, in turn, are physically connected to osteoblasts, surface-residing, bone-forming cells that share a common lineage with the osteocyte. This interconnectivity makes it possible for cells to sense biophysical signals and respond. For example, during walking as the leg bones are cyclically loaded and unloaded, the interstitial fluid flows back and forth across the osteocytes. In this manner, osteocytes are able to sense the global (macroscopic/organ) loading at the local (microscopic/cellular) level and many believe this phenomenon is critical to mechanotransduction. To simulate this in the laboratory osteocytes and osteoblasts (acting as young osteocytes) are plated on glass slides and inverted on parallel plate flow chambers that are used to subject groups of cells in monolayer to physiologic levels of fluid shear.

8.5.2   METHOD

To subject cells to oscillatory fluid flow, any loading platform can be modified. For example, using tubing connected on one end to a parallel plate inlet and on the other end to the syringe, an oscillatory waveform cycles the syringe plungers up and down which exposes the cells in the flow chamber to oscillatory fluid shear with a standard parabolic flow profile, Figure 8.30. The loading platform described in this text was utilized and a fixture to cycle the plungers while mounting the glass syringe to the base of the platform was machined.

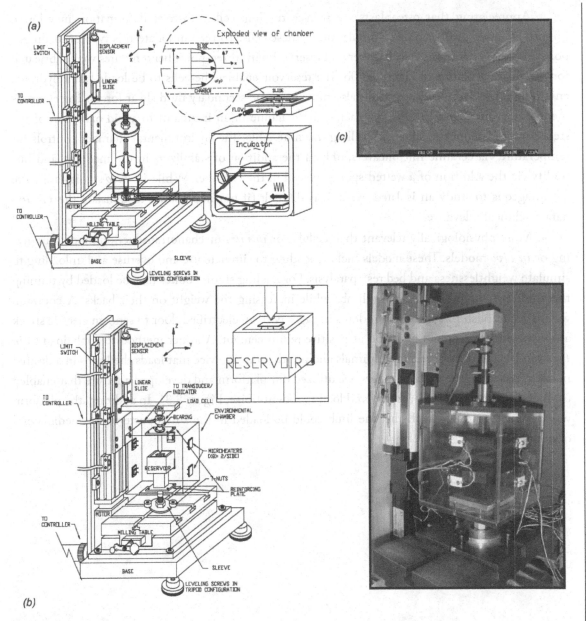

Figure 8.30: (a) Syringe assembly used to subject bone cells in a parallel plate flow chamber to physiologic levels of fluid shear. (b) Substrate deformation assembly used to subject osteoblasts to physiologic levels of bending (and secondary fluid shear). The environmental chamber was developed to maintain temperature; pH and humidity. (c) Cells in these systems are plated in monolayer on a synthetic substrate and subjected to stimulation.

Appreciating that osteoblasts are surface-residing cells, substrate deformation may be an important mode of physiologic stimulation in these cells. To accommodate this type of loading, a polycarbonate environmental chamber and reservoir with three-point bend fixtures were fabricated for the existing platform, Figure 8.30b. The reservoir enables the cells to be loaded in a hydrated environment to avoid cell death and also provides some secondary fluid shear forces. The environmental chamber is used in lieu of placing the machine in an incubator to avoid oxidation of the steel components of the slide and milling machine table. The environmental chamber controls for temperature via ceramic microheaters, pH via the addition of stabilizers in the medium, and humidity via the addition of a wetted sponge placed in the chamber. While *in vitro* studies are quite advantageous to study an isolated event, it is difficult (if not impossible) to translate the *in vitro* data to clinical relevance.

More physiologically relevant than cellular or *in vitro* mechanotransduction models are living or *in vivo* models. These models include loading to simulate use and overuse and unloading to simulate weightlessness and bed rest/paralysis. For loading studies bones may be loaded by training the animals to stand on their hindlimbs while increasing the weight on their backs. A common way to accomplish this is to train rodents in cages with an electrified floor to send an electric shock as negative reinforcement and food as positive reinforcement. Another method to apply load to *in vivo* models is to anesthetize the animals and put them in a device that loads the limbs in a desired manner. Based on the work of others, we utilized our platform and developed a fixture that enabled us to apply a concentrated, cantilevered load to rodent tibiae, Figure 8.31. In addition, the platform was designed to rotate such that the limb could be loaded in the anteroposterior or mediolateral orientation.

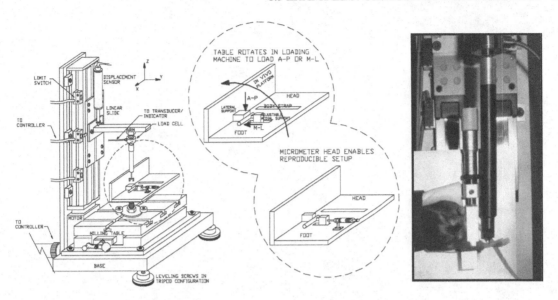

Figure 8.31: Bending assembly for *in vivo* stimulation studies. The platform rotates to load rodent tibiae in the anteroposterior (A–P; front-to-back) or mediolateral (M–L; side-to-side) orientations. A cantilevered loading is produced in which the proximal end of the tibia is held in place and the distal end is cyclically stimulated.

The field of Mechanobiology will continue to grow over the next several decades as we continue to work to understand how cells respond to mechanical environments and changes in those environments. As such, from a biomechanics standpoint this field is full of opportunities to develop new means of mechanical testing and testing techniques. Biomechanical engineers should realize the tremendous contributions they can make to the field of Mechanobiology and that the basics required in conducting quality biomechanics work are the same that are required in mechanobiology.

Bibliography

Aaron AD, Wiedel JD: Allograft use in orthopedic surgery. *Orthopedics* 1994;17:41-48.

Ajubi NE, Klein-Nulend J, Nijweide PJ, Vrijheidlammers T, Albas MJ, Burger EH. Pulsating fluid flow increases prostaglandin production by cultured chicken osteocytes: a cytoskeleton-dependent process. *Biochem Biophys Res Commun* 1996;225:62–68. DOI: 10.1006/bbrc.1996.1131.

Akhter MP, Cullen DM, Pedersen EA, Kimmel DB, Recker RR. Bone response to in vivo mechanical loading in two breeds of mice. *Calcif Tissue Int* 1998;63:442–449. DOI: 10.1007/s002239900554.

Akhter MP, Wells DJ, Short SJ, Cullen DM, Johnson ML, Haynatzki GR, Babij P, Allen KM, Yaworsky FB, Recker RR. Bone biomechanical properties in LRP5 mutant mice. *Bone* 2004;35:162–169. DOI: 10.1016/j.bone.2004.02.018.

Alverez JR, Gonzolez CC, Aranda RL, Blanco MF, Dehesa MC. Indications for use of the long gamma nail. *Clin Orthop* 1998;350:62–66.

An YH, Barfield WB, Draughn RA. Basic concepts of mechanical property measurement and bone biomechanics. In: *Mechanical testing of bone and the bone-implant interface*. New York: CRC Press; 2000, p23–40.

An YH, Bensen CV. General considerations of mechanical testing. In: *Mechanical testing of bone and the bone-implant interface*. New York: CRC Press; 2000, p119–132.

Aronsson DD, Carlson WE. Slipped capital femoral epiphysis. A prospective study of fixation with a single screw. *J Bone Joint Surg Am* 1992;74A: 810–819.

Aronsson, DD, Karol LA. Stable slipped capital femoral epiphysis: Evaluation and management. *J Am Acad Orthop Surg* 1996;4: 173–181.

Ascenzi MG, Benvenuti A, Ascenzi A. Single osteon micromechanical testing. In: *Mechanical testing of bone and the bone-implant interface*. New York: CRC Press; 2000, p271–290.

Barink M. (University of Nijmegen – Orthopaedic Research Lab). Mechanical testing of sawbones 3rd generation composite femora. In-house report, 2001.

Benevenia J, Zimmerman M, Keating J. Mechanical environment affects allograft incorporation. *J Biomed Mater Res* 2000;53(1): 67–72. DOI: 10.1002/(SICI)1097-4636(2000)53:1<67::AID-JBM9>3.0.CO;2-#.

Bennett GL, Cameron B, Njus G, Saunders M, Kay DB. Tibiotalocalcaneal arthrodesis: a biomechanical assessment of stability. *Foot Ankle Int* 2005;26(7):530–536.

Bergandi JA, Saunders MM, Kaag M, Jacobs CR, Parrish WM. Torsional and bending evaluation of femoral allograft fixation techniques in a cadaveric model: A comparison of double plate to IM nail and plate fixation. *Trans Orthop Res Soc*, 2001.

Black KP, Saunders MM, Stube KC, Moulton M, Jacobs CR: Effects of interference screw length on tibial tunnel fixation for anterior cruciate ligament reconstruction. *Am J Sports Med* 2000;28:846–849.

Black KP, Saunders MM. Expansion anchors for use in anterior cruciate ligament reconstruction: establishing proof of concept in a benchtop analysis. *Med Eng Phys* 2005;27:425–434. DOI: 10.1016/j.medengphy.2004.11.004.

Blanco JS, Taylor B, Johnston CE II. Comparison of single pin versus multiple pin fixation in treatment of slipped capital femoral epiphysis. *J Pediatr Orthop* 1992;12: 384–389. DOI: 10.1097/01241398-199205000-00019.

Bonewald LF. Osteocytes as dynamic multifunctional cells. *Ann NY Acad Sci* 2007;1116:281–290. DOI: 10.1196/annals.1402.018.

Bottlang M, Simnacher M, Schmidt H, Brand RA, Claes L. A cell strain system for small homogenous strain applications. *Biomed Tech* (Berl) 1997;42:305–309. DOI: 10.1515/bmte.1997.42.11.305.

Brien EW, Terek RM, Healy JH, Lane JM. Allograft reconstruction after proximal tibial resection for bone tumors. An analysis of function and outcome comparing allograft and prosthetic reconstruction. *Clin Orthop* 1994;303: 116–127.

Brodt MD, Ellis CB, Silva MJ. Growing C57B1/6 mice increase whole bone mechanical properties by increasing geometric and material properties. *J Bone Miner Res* 1999;14(12):2159–2166. DOI: 10.1359/jbmr.1999.14.12.2159.

Brown CH, Aaron TH, Hipp JA, Myers ER, Wilson HC. The biomechanics of interference screw fixation of patellar tendon anterior cruciate ligament grafts. *Am J of Sports Med* 1993;21(6):880–886. DOI: 10.1177/036354659302100622.

Brown TD, Bottlang M, Pedersen DR, Banes AJ. Loading paradigms – intentional and unintentional – for cell culture mechanostimulus. *Am J Med Sci* 1998;316:162–168. DOI: 10.1097/00000441-199809000-00003.

Brown TD. Techniques for mechanical stimulation of cells in vitro: a review. *J Biomech* 2000;33(1):3–14. DOI: 10.1016/S0021-9290(99)00177-3.

Buckwalter JA, Einhorn TA, Simon S. *Orthopaedic basic science* 2nd ed. Rosemont: AAOS; 2000.

Caler WE, Carter DR. Bone creep-fatigue damage accumulation. *J Biomech* 1989;22:625–635. DOI: 10.1016/0021-9290(89)90013-4.

Carter DR, Beaupre GS, Giori NJ, Helms JA. Mechanobiology of skeletal regeneration. *Clin Orthop* 1998;S355:S41–55. DOI: 10.1097/00003086-199810001-00006.

Cech O, Sosna A. Principles of the surgical treatment of subtrochanteric fractures. *Orthop Clin North Am* 1974;5(3):651–662.

Courpied J, Watin-Augouard L, Postel M. Fractures du femur chez les sujets porteurs de prostheses totales de hanche ou de genou. *Int Orthop* 1987;11: 109–115. DOI: 10.1007/BF00266695.

Cowin SC. Bone stress adaptation models. *J Biomech* Eng 1993;115:528–533. DOI: 10.1115/1.2895535.

Cowin SC, Weinbaum S. Strain amplification in the bone mechanosensory system. *Am J Med Sci* 1998;316:184–188. DOI: 10.1097/00000441-199809000-00006.

Cowin SC. Bone poroelasticity. In: *Bone mechanics handbook*. CRC Press, Boca Raton, Florida, Chapter 23:1, 2000.

Currey J. Measurement of the mechanical properties of bone. A recent history. *Clin Orthop* 2009;467:1948–1954. DOI: 10.1007/s11999-009-0784-z.

Czitrom A, Gross A. *Allografts in Orthopaedic Practice*. Baltimore: Williams and Wilkins; 1992.

Dempster WT, Liddicoat R. Compact bone as a non-isotropic material. *Am J Anat* 1952;91:331–362. DOI: 10.1002/aja.1000910302.

DiMaio FR, Haher TR, Splain SH, Mani VJ. Stress-riser fractures of the hip after sliding screw plate fixation. *Orthop Rev* 1992;21(10): 1229–1231.

Donahue HJ, Chen Q, Jacobs CR, Saunders MM, Yellowley CE. Bone cells and mechanotransduction. In: *Molecular biology in orthopaedics*. Rosemont: AAOS; 2003, 179–190.

Draper ERC, Goodship AE. A novel technique for four-point bending of small bone samples with semi-automatic analysis. *J Biomech*, 2003; 36(10):1497–1502. DOI: 10.1016/S0021-9290(03)00129-5.

Early SD, Hedman TP, Reynolds RAK: Biomechanical analysis of compression screw fixation versus in situ pinning in slipped capital femoral epiphysis. *J Pediatr Orthop* 2001;21: 183–188. DOI: 10.1097/01241398-200103000-00010.

Einhorn TA, Wakley GK, Linkhart S, Rush EB, Maloney S, Faierman E, Baylink DJ. Incorporation of sodium fluoride into cortical bone does not impair the mechanical properties of the appendicular skeleton of rats. *Calcif Tissue Int* 1992;51:127–131. DOI: 10.1007/BF00298500.

Evans FG. *Mechanical properties of bone*. Springfield: Thomas; 1973.

Fielding WJ, Cochran GVB, Zickel RE. Biomechanical characteristics and surgical management of subtrochanteric fractures. *Orthop Clin North Am* 1974;5(3):629–650.

Forwood, MR, Parker AW. Microdamage in response to repetitive torsional loading in the rat tibia. *Calcif Tissue Int* 1989;45:47–53. DOI: 10.1007/BF02556660.

Frangos JA, Eskin SG, McIntire LV, Ives CL. Flow effects on prostacyclin production by cultured human endothelial cells. *Science* 1985;227:1477–1479. DOI: 10.1126/science.3883488.

Frangos JA, McIntire LV, Eskin SG. Shear stress induced stimulation of mammalian cell metabolism. *Biotechnol Bioeng* 1988;32:1053–1060.

Fritton JC, Myers ER, van der Muelen MCH, Bostrom MPG, Wright TM. Validation of a loading apparatus: Characterization of murine tibial surface strains in vivo. *Trans Orthop Res Soc*, 2001.

Froimson AI. Treatment of comminuted subtrochanteric fractures. *Surg Gynecol Obstet* 1970;131:465–472.

Fulkerson JP, Cautilli R, Hosick WB, Wright J. Divergence angles and their effect on the fixation strength of the Kurosaka screw. Presented at the Jefferson Orthopaedic Society Meeting, Philadelphia, November, 1991.

Furman BR, Saha S. Torsional testing of bone. In: *Mechanical testing of bone and the bone-implant interface*. New York: CRC Press; 2000, p219–232.

Garbuz DS, Masri BA, Dunjcan CP. Periprosthetic fractures of the femur: Principles of prevention and management. In: *Instructional course lectures*, Rosemont: AAOS; 1998, p237–242.

Gere JM, Timoshenko SP. *Mechanics of materials*, 2nd ed. California: Brooks/Cole Engineering Division; 1984.

Goodman WW, Johnson JT, Robertson WW Jr. Single screw fixation for acute and acute-on-chronic slipped capital femoral epiphysis. *Clin Orthop* 1996;322:86–90. DOI: 10.1097/00003086-199601000-00010.

Gross TS, Srinivasan S, Liu CC, Clemens TL, Bain SD. Non-invasive loading of the murine tibia: an in vivo model for the study of mechanotransduction. *J Bone Miner Res* 2002;17(3):493–501. DOI: 10.1359/jbmr.2002.17.3.493.

Guo XE. Mechanical properties of cortical bone and cancellous bone tissue. In: *Bone Mechanics Handbook*, Florida: CRC Press; 2000, Chapter 10:1–21.

Hanson GW, Tullos HS. Subtrochanteric fractures of the femur treated with nail-plate devices: a retrospective study. *Clin Orthop* 1978;131:191–194.

Harris T, Ruth JT, Szivek J, Haywood B. The effect of implant overlap on the mechanical properties of the femur. *J Trauma* 2003;54:930–935. DOI: 10.1097/01.TA.0000060999.54287.39.

Heiple KG, Brooks DB, Sampson BL, Burstein AH. A fluted intramedullary rod for subtrochanteric fractures. Biomechanical considerations and preliminary clinical results. *J Bone Joint Surg Am* 1979;61:730–737.

Herman MJ, Dormans JP, Davidson RS, Drummond DS, Gregg JR. Screw fixation of grade III slipped capital femoral epiphysis. *Clin Orthop* 1996;322:77–85. DOI: 10.1097/00003086-199601000-00009.

Hillam RA, Skerry TM. Inhibition of bone resorption and stimulation of formation by mechanical loading of the modeling rat ulna in vivo. *J Bone Miner Res* 1995;10(5):683–689. DOI: 10.1002/jbmr.5650100503.

Hornicek F, Gebhardt M, Tomford W. Factors affecting nonunion of the allograft host junction. *Clin Orthop* 2001;382:87–98. DOI: 10.1097/00003086-200101000-00014.

Huang-Brown KM, Saunders MM, Kirsch T, Donahue HJ, Reid JS. The effect of COX-2 specific inhibitors on fracture healing in the adult rat femur. *J Bone Joint Surg Am* 2004;86-A(1):116–123.

Hulstyn M, Fadale PD, Abate J, and Walsh WR. Biomechanical evaluation of interference screw fixation in a bovine patellar bone-tendon-bone autograft complex for anterior cruciate ligament reconstruction. *Arthroscopy* 1993;9(4):417–424. DOI: 10.1016/S0749-8063(05)80316-0.

Hung CT, Pollack SR, Reilly TM, Brighton CT. Real-time calcium response of cultured bone cells to fluid flow. *Clin Orthop* 1995;313:256–269.

Hung CT, Allen FD, Pollack SR, Brighton CT. Intracellular Ca2+ stores and extracellular Ca2+ are required in the realtime Ca2+ response of bone cells experiencing fluid flow. *J Biomech* 1996;29:1411–1417. DOI: 10.1016/0021-9290(96)84536-2.

Jacobs CR, Yellowley CE, Davis BR, Zhou Z, Donahue HJ. Differential effect of steady versus oscillating flow on bone cells. *J Biomech* 1998;31:969–976. DOI: 10.1016/S0021-9290(98)00114-6.

Jones DB, Broeckmann E, Pohl T, Smith EL. Development of a mechanical testing and loading system for trabecular bone studies for long term culture. *Eur Cell Mater* 2003;5:48–60.

Karol LA, Doane RM, Cornicelli SF, Zak PA, Haut RC, Manoli A II. Single versus double screw fixation for treatment of slipped capital femoral epiphysis: A biomechanical analysis. *J Pediatr Orthop* 1992;12:741–745. DOI: 10.1097/01241398-199211000-00008.

Kashiwagi D. Intra-articular changes of the osteoarthritic elbow, especially about the fossa olecrani. *J Jpn Orthop Assoc* 1978;52:1376–1382.

Kashiwagi D. Osteoarthritis of the elbow joint: intra-articular changes and the special operative procedure. Outerbridge-Kashiwagi method. In: *Elbow joint*. New York: Elsevier; 1985, p177–188.

Kashiwagi D. Osteoarthritis of the elbow joint. In: *Elbow joint*. Procs International Congress. Amsterdam: Elsevier; 1986.

Kato Y, Windle JJ, Koop BA, Mundy GR, Bonewald LF. Establishment of an osteocyte-like cell line, MLO-Y4. *J Bone Miner Res* 1997;12:2014–2023. DOI: 10.1359/jbmr.1997.12.12.2014.

Kibbe RR, Meyer RO, Neely JE, White WT. *Machine tool practices* 8th ed. Upper Saddle River: Pearson Prentice Hall; 2002.

Kibiloski LJ, Doane RM, Karol LA, Haut RC, Loder RT. Biomechanical analysis of single versus double-screw fixation in slipped capital femoral epiphysis at physiological load levels. *J Pediatr Orthop* 1994;14:627–630. DOI: 10.1097/01241398-199409000-00015.

Klein-Nulend, J, Burger EH, Semeins CM, Raisz LG, Pilbeam CC. Pulsating fluid flow stimulates prostaglandin release and inducible prostaglandin G/H synthase mRNA expression in primary mouse bone cells. *J Bone Miner Res* 1997;12:45–51. DOI: 10.1359/jbmr.1997.12.1.45.

Kohn D, Rose C. Primary stability of interference screw fixation. Influence of screw diameter and insertion torque. *Am J Sports Med* 1994;22:334–338. DOI: 10.1177/036354659402200307.

Kraemer WJ, Hearn TC, Powell JN, Mahomed N. Fixation of segmental subtrochanteric fractures. A biomechanical study. *Clin Orthop* 1996;332:71–79. DOI: 10.1097/00003086-199611000-00010.

Kruger DM, Herzenberg JE, Viviano DM, Hak DJ, Goldstein SA. Biomechanical comparison of single- and double-pin fixation for acute slipped capital femoral epiphysis. *Clin Orthop* 1990;259:277–281.

Kurosaka M, Yoshiya S, Andrish JT. A biochemical comparison of different surgical techniques of graft fixation in anterior cruciate ligament reconstruction. *Am J Sports Med* 1987;15(3):225–229. DOI: 10.1177/036354658701500306.

Lanyon LE. Osteocytes, strain detection, bone modeling and remodeling. *Calcif Tissue Int* 1993; 53(Suppl):102–107. DOI: 10.1007/BF01673415.

Larson JE, Chao EY, Fitzgerald RH. Bypassing femoral cortical defects with cemented intramedullary stems. *J Orthop Res* 1991;9:414–421. DOI: 10.1002/jor.1100090314.

Levenston ME, Beaupre GS, van der Meulen MC. Improved method for analysis of whole bone torsion tests. *J Bone Miner Res* 1994;9(9):1459–1465. DOI: 10.1002/jbmr.5650090919.

Levy RN, Siegel M, Sedlin ED, Siffert RS. Complications of Ender-pin fixation of basicervical, intertrochanteric, and subtrochanteric fractures of the hip. *J Bone Joint Surg Am* 1983;65:66–69.

Lind PM, Lind L, Larsson S, Orberg J. Torsional testing and peripheral quantitative computed tomography in rat humerus. *Bone* 2001;29(3):265–270. DOI: 10.1016/S8756-3282(01)00576-2.

Loder RT: Unstable slipped capital femoral epiphysis. *J Pediatr Orthop* 2001;21:694–699. DOI: 10.1097/01241398-200109000-00027.

Lopez MJ, Markel MD. Bending tests of bone. In: *Mechanical testing of bone and the bone–implant interface*. New York: CRC Press; 2000, p207–218.

Lundy DW, Acevedo JI, Ganey TM, Ogden JA, Hutton WC. Mechanical comparisons of plates used in the treatment of stable subtrochanteric femur fractures. *J Orthop Trauma* 1999;13:534–538. DOI: 10.1097/00005131-199911000-00003.

Mabrey J. Periprosthetic Fractures. In: *Fractures in Adults*. Vol. 1. Philadelphia: Lippincott/Rave; 1996, p585–586.

Mahomed N, Harrington I, Kellam J, Maistrelli G, Hearn T, Vroemen J. Biomechanical analysis of the gamma nail and sliding hip screw. *Clin Orthop* 1994;304:280–288.

Mankin H, Gebhardt M, Jennings C. Long term results of allograft replacement in the management of bone tumors. *Clin Orthop* 1996;24:86–97. DOI: 10.1097/00003086-199603000-00011.

Martin RB, Burr CD, Sharkley NA. Skeletal Biology. In: *Skeletal tissue mechanics*. New York: Springer-Verlag; 1998, p32–33. DOI: 10.1007/978-1-4757-2968-9_2.

Matthews LS, Lawrence SJ, Yahiro MA, Sinclair MR. Fixation strengths of patellar tendon-bone grafts. *Arthroscopy* 1993;9(1):76–81. DOI: 10.1016/S0749-8063(05)80348-2.

Mikic B, van der Muelen MCH, Kingsley DM, Carter DR. Long bone geometry and strength in adult BMP-5 deficient mice. *Bone* 1995;16:445–454.

Mikic B, Battaglia TC, Taylor EA, Clark RT. The effect of growth/differentiation factor-5 deficiency on femoral composition and mechanical behavior in mice. *Bone* 2002;30(5):733–737. DOI: 10.1016/S8756-3282(02)00699-3.

Miller CM, Tibone JE, Hewitt M, Kharrazi FD, Elattrache NS. Interference screw divergence in femoral tunnel fixation during endoscopic anterior cruciate ligament reconstruction using hamstring grafts. *Arthroscopy* 2002;18(5):510–514. DOI: 10.1053/jars.2002.30653.

Moed BR, Watson JT. Retrograde intramedullary nailing, without reaming, of fractures on the femoral shaft in multiply injured patients. *J Bone Joint Surg Am* 1995;77(10):1520–1527.

Morrey BF. Primary degenerative arthritis of the elbow. *J Bone Joint Surg Br* 1992;74B:409–413.

Muir P, Johnson K. Tibial intercalary allograft incorporation: Comparison and fixation with locked intramedullary nail and dynamic compression plate. *J Orthop Res* 1995;13(1):132–137. DOI: 10.1002/jor.1100130119.

Muller M, Schneider R. *Manual of internal fixation techniques recommended by the AO-ASIF group.* 3rd ed. Berlin: Springer; 1991. DOI: 10.1007/978-3-662-02695-3.

Nazarian A, Bauernschmitt M, Eberle C, Meier D, Müller R, Snyder BD. Design and validation of a testing system to assess torsional cancellous bone failure in conjunction with time-lapsed micro-computed tomographic imaging. *J Biomech* 2008;41:3496–3501. DOI: 10.1016/j.jbiomech.2008.09.014.

Nazarian A, Entezari V, Vartanians V, Muller R, Snyder BD. An improved method to assess torsional properties of rodent long bones. *J Biomech* 2009;42:1720–1725. DOI: 10.1016/j.jbiomech.2009.04.019.

Ostrum RF, DiCicco J, Lakatos R, Poka A. Retrograde intramedullary nailing of femoral diaphyseal fractures. *J Orthop Trauma* 1998;12:464–468. DOI: 10.1097/00005131-199809000-00006.

Ostrum RF, Agarwal A, Lakatos R, Poka A. Prospective comparison of retrograde and antegrade femoral intramedullary nailing. *J Orthop Trauma* 2000;14:496–501. DOI: 10.1097/00005131-200009000-00006.

Owan I, Burr DB, Turner CH, Qui J, Tu Y, Onyia JE, Duncan RL. Mechanotransduction in bone: Osteoblasts are more responsive to fluid forces than mechanical strain. *Am J Physiol Cell Physiol* 1997;273(3 Pt 1):C810–815.

Pankovich AM, Tarabishy IE. Ender nailing of the intertrochanteric and subtrochanteric fractures of the femur. *J Bone Joint Surg Am* 1980;62:635–645.

Patterson BM, Routt MLC Jr, Benirschke SK, Hansen ST Jr. Retrograde nailing of femoral shaft fractures. *J Trauma* 1995;38(1):38–43. DOI: 10.1097/00005373-199501000-00012.

Paul JP. Approaches to design. Force actions transmitted by joints in the human body. *Proc R Soc Lond B Biol Sci* 1976;192:163–172. DOI: 10.1098/rspb.1976.0004.

Pelker RR, Friedlaender GE, Markham TC, Panjabi MM, Moen CJ. Effects of freezing and freeze-drying on the biomechanical properties of rat bone. *J Orthop Res* 1984;1:405–411. DOI: 10.1002/jor.1100010409.

Petrey JS, Saunders MM, Kluemper GT, Beeman CS. Temporary anchorage device insertion variables: effects on retention. *Angle Orthod* 2010;80(4):446–453. DOI: 10.2319/070309-376.1.

Piekarski K, Munro M. Transport mechanism operating between blood supply and osteocytes in long bones. *Nature* 1977;269:80–82. DOI: 10.1038/269080a0.

Pomeroy G, Baltz M, Pierz K, Nowak M, Post W, Fulkerson JP. The effects of bone plug length and screw diameter on the holding strength of bone-tendon-bone grafts. *Arthroscopy* 1998;14(2):148–152. DOI: 10.1016/S0749-8063(98)70033-7.

Pritchett J, Perdue K. Mechanical factors in slipped capital femoral epiphysis. *J Pediatr Orthop* 1988;8:385–388. DOI: 10.1097/01241398-198807000-00001.

Pugh KJ, Morgan RA, Gorczyca JT, Pienkowski D. A mechanical comparison of subtrochanteric fracture fixation. *J Orthop Trauma* 1998;12:324–329. DOI: 10.1097/00005131-199806000-00005.

Resnick AM. Optimizing interference fixation for cruciate ligament reconstruction. *Trans Orthop Res Soc* 1990;15:519.

Ricci WM, Bellabarba C, Evanoff B, Herscovici D, DiPasquale T, Sanders R. Retrograde versus antegrade nailing of femoral shaft fractures. *J Orthop Trauma* 2001;15:161–169. DOI: 10.1097/00005131-200103000-00003.

Robling AG, Turner CH. Mechanotransduction in bone: Genetic effects on mechanosensitivity in mice. *Bone* 2002;31(5):562–569. DOI: 10.1016/S8756-3282(02)00871-2.

Rubin, J, Fan X, Biskobing DM, Taylor WR, Rubin CT. Osteoclastogenesis is repressed by mechanical strain in an in vitro model. *J Orthop Res* 1999;17(5):639–645.

Rubin J, Murphy T, Nanes MS, Fan X. Mechanical strain inhibits expression of osteoclast differentiation by murine stromal cells. *Am J Physiol Cell Physiol* 2000;278(6):C1126–1132.

Russell TA, Taylor JC. Subtrochanteric fractures in the femur. In: *Skeletal Trauma*. Philadelphia: WB Saunders; 1992, p1485–1524.

Sanders R, Koval KJ, DiPasquale T, Helfet DL, Frankle M. Retrograde reamed femoral nailing. *J Orthop Trauma* 1993;7(4):293–302. DOI: 10.1097/00005131-199308000-00001.

Saunders MM, You J, Trosko JE, Yamasaki H, Donahue HJ, Jacobs CR. Gap junctions and fluid flow response in MC3T3-E1 cells. *Am J Physiol Cell Physiol* 2001;281(6):1917–1925.

Saunders MM, You J, Zhou Z, Li Z, Yellowley CE, Kunze E, Jacobs CR, Donahue HJ. Fluid-flow induced prostaglandin E2 response of osteoblastic ROS 17/2.8 cells is gap junction-mediated and independent of cytosolic calcium. *Bone* 2003;32:350–356. DOI: 10.1016/S8756-3282(03)00025-5.

Saunders MM, Donahue, HJ. Development of a cost-effective loading machine for biomechanical analysis of mouse transgenic models. *Med Eng Phys* 2004;26:595–603. DOI: 10.1016/j.medengphy.2004.04.005.

Saunders MM, Taylor AF, Du C, Zhou Z, Pellegrini VD Jr, Donahue H. Mechanical stimulation effects on functional end effectors in osteoblastic MG-63 cells. *J Biomech* 2006;39(8):1419–1427. DOI: 10.1016/j.jbiomech.2005.04.011.

Saunders MM, Lee J. The influence of mechanical environment on bone healing and distraction osteogenesis. In: *Atlas of the oral and maxillofacial surgery clinics of North America* (devoted to distraction). Philadelphia: Elsevier; 2008, p147–158.

Saunders M, Baxter C, Abou-Elella A, Kunselman AR, Trussell J. BioGlue and dermabond save time and leak less than sutured microsurgical vasovastostomy. *Fertil Steril* 2008;91(2):560–565. DOI: 10.1016/j.fertnstert.2007.12.006.

Saunders MM, Burger RB, Kalantari B, Nichols AD, Witman C. Development of a cost-effective torsional unit for small-scale biomechanical testing. *Med Eng Phys* 2010;32:802–807. DOI: 10.1016/j.medengphy.2010.05.004.

Saunders MM. Small-scale mechanical testing: Applications to bone biomechanics and mechanobiology. In: *Time dependent Constitutive behavior and fracture/failure processes*, Vol 3. NY: Springer; 2011, p345–352. DOI: 10.1007/978-1-4419-9794-4_48.

Saunders MM. Biomimetics in bone cell mechanotransduction: Understanding bone's response to mechanical loading. In: *Advances in biomimetics*. Intech (open access); 2011, p317–348.

Savoie, FH, Nunley, PD and Field, LD. Arthroscopic management of the arthritic elbow: Indications, techniques, and results. *J Shoulder Elbow Surg* 1999;8:214–219. DOI: 10.1016/S1058-2746(99)90131-3.

Segal LS, Jacobson JA, Saunders MM. Biomechanical analysis of in situ single versus double screw fixation in a non-reduced slipped capital femoral epiphysis model. *J Pediatr Orthop* 2006;26(4):479–485. DOI: 10.1097/01.bpo.0000226285.46943.ea.

Silva MJ, Ulrich SR. In vitro sodium exposure decreases torsional and bending strength and increases ductility of mouse femora. *J Biomech* 2000;33:231–234. DOI: 10.1016/S0021-9290(99)00158-X.

Swiontkowski MF, Hansen ST Jr, Kellam J. Ipsilateral fractures of the femoral neck and shaft. *J Bone Joint Surg Am* 1984;66(2):260–268.

Taylor AF, Saunders MM, Shingle D, Cimbala JM, Zhou Z, Donahue HJ. Osteocytes communicate fluid flow-mediated effects to osteoblasts altering phenotype. *Am J Physiol Cell Physiol* 2007;292:545–552. DOI: 10.1152/ajpcell.00611.2005.

Teitelbaum SL. Bone resorption by osteoclasts, *Science* 2000;289(5484):1504–1508. DOI: 10.1126/science.289.5484.1504.

Tencer AF, Johnson KD, Johnston DW, Gill K. A biomechanical comparison of various methods of stabilization of subtrochanteric fractures of the femur. *J Orthop Res* 1984;2:297–305. DOI: 10.1002/jor.1100020312.

Tencer A, Johnson K, Kyle R, Fu F. Biomechanics of fracture and fracture fixation. In: *Instructional course lectures.* Rosemont: AAOS; 1993, p19–55.

Tomita F, Yasuda K, Mikami S, Sakai T, Yamazaki S, Tohyama H. Comparisons of intraosseous graft healing between the doubled flexor tendon graft and the bone-patellar tendon-bone graft in anterior cruciate ligament reconstruction. Arthroscopy 2001;17(5):461–476. DOI: 10.1053/jars.2001.24059.

Tornetta P III, Tiburzi D. Antegrade or retrograde reamed femoral nailing: a prospective, randomized trial. *J Bone Joint Surg Br* 2000;82:652–654. DOI: 10.1302/0301-620X.82B5.10038.

Turner CH, Akhter MP, Raab DM, Kimmel DB, Recker RR. A noninvasive in vivo model for studying strain adaptive bone remodeling. *Bone* 1991;12:73–79. DOI: 10.1016/8756-3282(91)90003-2.

Turner CH, Burr DB. Basic biomechanical measurements of bone: A tutorial. *Bone* 1993;14:595–608. DOI: 10.1016/8756-3282(93)90081-K.

Turner CH. Bone strength: Current concepts. *Ann NY Acad Sci* 2006;1068:429–446. DOI: 10.1196/annals.1346.039.

Vaananen K, Jalovaara P, Lepola V. The effect of immobilization on the torsional strength of the rat tibia. *Clin Orthop* 1993;297:55–61.

Van Rietbergen B, Huiskes R. Elastic constants of cancellous bone. In: *Bone mechanics handbook.* Florida: CRC Press; 2000; p15–19.

Walker JR. *Machining fundamentals.* Tinley Park: Goodheart-Wilcox Company; 2000.

Ward WT, Stefko J, Wood KB, Stanitski CL. Fixation with a single screw for slipped capital femoral epiphysis. *J Bone Joint Surg Am* 1992;74A:799–809.

Weinbaum S, Cowin SC, Zeng Y. A model for the excitation of osteocytes by mechanical loading-induced bone fluid shear stresses. *J Biomech* 1994;3:339–360. DOI: 10.1016/0021-9290(94)90010-8.

Wiss DA, Sima W, Brien WW. Ipsilateral fractures of the femoral neck and shaft. *J Orthop Trauma* 1992;6(2):159–166. DOI: 10.1097/00005131-199206000-00005.

Wu CC, Shih CH, Lee ZL. Subtrochanteric fractures treated with interlocking nails. *J Trauma* 1991;31:326–333. DOI: 10.1097/00005373-199103000-00004.

Yang W, Li Z, Shi W, Xie B, Yang M. Review on auxetic materials. *J Mater Sci* 2004;39(10):3269–3279. DOI: 10.1023/B:JMSC.0000026928.93231.e0.

Zickel RE. A new fixation device for subtrochanteric fractures of the femur: a preliminary report. *Clin Orthop* 1967;54:115–123. DOI: 10.1097/00003086-196709000-00013.

Zickel RE. An intramedullary fixation device for the proximal part of the femur. *J Bone Joint Surg Am* 1976;58:866–872.

Author Biography

Dr. Marnie Saunders is a native of northeast Ohio. She received her B.S. in Mechanical Engineering from The University of Akron in 1991. She completed graduate education at The University of Akron in Engineering (Biomechanics), receiving her M.S. and Ph.D. degrees in 1994 and 1998, respectively. Under the guidance of the late Dr. Glen O Njus, her graduate research was focused in orthopaedic biomechanics and lower limb prosthetics.

In 1998, Dr. Saunders accepted a two-year post-doctoral fellowship at the Pennsylvania State University College of Medicine in Hershey, PA, in the Department of Orthopaedics and Rehabilitation. She split her research efforts between orthopaedic biomechanics and bone cell mechanobiology. Upon completion of her fellowship, she joined the faculty at PSU and remained there for six years working in the Department of Orthopaedics and Rehabilitation conducting research.

Presented with the opportunity to teach graduate students, Dr. Saunders accepted a position in the Center for Biomedical Engineering at the University of Kentucky. She continued her research in bone cell mechanobiology and had the opportunity to transition from orthopaedic biomechanics to craniofacial biomechanics.

In 2010 she was given the opportunity to return to her alma mater, The University of Akron, where she is currently an Associate Professor in the Department of Biomedical Engineering. She teaches undergraduate courses in Freshman Design, Musculoskeletal Tissue Mechanics and graduate courses in Bone Biomechanics and Mechanobiology and Continuum Mechanics. She continues her interest and work in design and fabrication of cost-effective, small-scale biomechanical testing platforms, orthopaedic biomechanics, and mechanobiology. Currently her interest is in combining biomimicry and mechanobiology. Her research group is developing *in vitro* bone cell models that attempt to incorporate key physiologic features to improve research relevance. The models are being used to study how bone cells (osteocytes, osteoclasts and osteoblasts) coordinate their activity in response to alterations in loading environments (microgravity to microdamage).

Dr. Saunders has had over $3 million in research support and her research has been funded by The Whitaker Foundation, The National Institutes of Health and The National Science Foundation. She is the recipient of prestigious awards including the NIH NIA K25 Career Development

Award and an NSF Career Award. Her current funding includes NSF (Career) and NIH NIDCR (AREA) funding. She is an author/co-author on several research awards (ASBMR, BMES, AAOS, AOA) and while at PSU she received the 2004 Hinkle Society (Rising Faculty Star) Award and a 2005 faculty Mentor Award (Department of Mechanical Engineering). She is an author/co-author on more than 45 book chapters and research publications and over 150 international, national and local research presentations. In 2014, the work of her research group was featured in the journal of *International Innovation* and she has recently accepted the invitation to serve as guest editor of a biomimicry textbook for Springer Publishing.

Printed in the United States
by Baker & Taylor Publisher Services

Printed in the United States
by Baker & Taylor Publisher Services